DATE DUE

Tropical Diseases Research Series
No. 1

Role of the Spleen
in the Immunology of
Parasitic Diseases

Proceedings of the Meeting held in
Geneva 12–14 June 1978

Published on behalf of the

UNDP/WORLD BANK/WHO
Special Programme for
Research and Training in Tropical Diseases

Schwabe & Co. AG, Basel

This book contains the views and statements of
an international group of experts meeting under
the auspices of the UNDP/WORLD BANK/WHO Special
Programme for Research and Training in Tropical
Diseases. Authors alone are responsible for
views expressed in signed chapters.

© 1979 by Schwabe & Co. AG, Basel
ISBN 3-7965-0751-4

Preface

The Tropical Diseases Research Series is a publication of the UNDP/WORLD BANK/WHO Special Programme for Research and Training in Tropical Diseases. The series will deal with current outstanding problems in research related to the control of major infectious tropical diseases. It is intended to attract scientists working in other disciplines to the problems and opportunities for research which are offered by the infective agents of concern to the Programme, and to acquaint those already working on these organisms with new awareness for investigation which advances in other biomedical sciences may offer.

The Special Programme is a goal-oriented research and training programme with two objectives:

- Research and development to obtain new and improved tools for the control of major tropical diseases, and

- Strengthening of the research capability of the tropical countries.

The research is conducted on a global basis by multidisciplinary Scientific Working Groups; the training and institution strengthening activities are limited to the tropical countries where the diseases are endemic.

The six diseases initially selected for attack are: malaria, schistosomiasis, filariasis (including onchocerciasis), trypanosomiasis (both African and the American form called Chagas' disease), leishmaniasis and leprosy. Scientific Working Groups are also active in trans-disease areas: Biomedical Sciences, Vector Control, Epidemiology, and Social and Economic Research.

Scientists interested in participating in the Special Programme are invited to write for further information.

Dr Adetokunbo O. Lucas, Director
Special Programme for Research
    and Training in Tropical Diseases
c/o World Health Organization
Geneva, Switzerland

Contents

Introduction

J.H. Humphrey, Department of Immunology,
Royal Postgraduate Medical School, Hammersmith Hospital,
London, England.

That the spleen is important for enabling the body to cope
with bacterial infections is witnessed by the greatly increased
liability to septicaemia of persons whose spleen has been removed,
for one reason or another. The subject of the discussion which
follows is to be how far the spleen is important as a major or
unique site for development and effective function of immunity in
a variety of parasitic diseases. It may well be the first time
ever that a group of persons with such widely differing backgrounds
has met to focus their attention on these particular questions,
but the role of the spleen has intrigued persons with an enquiring
mind for a very long time. By way of introduction, I have chosen
a quotation from a book by Magnus Falconer published in 1777,
describing the observations and experiments of the late
Mr William Hewson, FRS. The book describes "The Particles of the
Blood in the Human Subject and in other Animals; with an account of
the Structure and Offices of the Lymphatic Glands, of the Thymus
Gland, and of the Spleen". Chapter IV "On the Situation and
Structure of the Spleen" begins as follows:

> "It hath at all times been a matter of surprise among
> the learned, that a viscus so large, and so advantageously
> situated as the Spleen is, added to the frequent
> opportunities of inspecting it in different states of
> health, should, notwithstanding, have its uses so
> involved in obscurity as to elude the researches of so
> many ingenious and industrious inquirers.

Not that the Spleen has at any time been considered
as useless, for at different periods a variety of
different offices have been assigned to it. Among
the ancients, the most celebrated opinion was, that
it made the atra bilis or succus melancholicus, which
they supposed was carried by the basa brevia into the
stomach: but later observations have entirely
exploded that idea, in so much that the very term is
almost extinct. And we shall endeavour to prove,
that the more modern opinion, of its producing some
change on the blood, preparatory to the secretion of
bile, hath no better foundation in nature.

But it will be unnecessary to repeat the various
opinions that have been entertained at different
times respecting the use of this viscus. Our present
endeavour will be to describe its situation and
structure and afterwards to inquire into some
particulars respecting its use."

I should add that Mr Hewson proved to his own satisfaction
that the spleen was the site of production of red corpuscles of the
blood – even though it was known in his time that dogs whose spleen
had been removed appeared to suffer no ill effects from the
operation! Nevertheless the sentiments expressed in the last
paragraph of the above quotation seem entirely appropriate to the
discussion which is to follow.

Many important advances in our understanding of the nature and
control of immune responses have come from studies made using
dispersed cell suspensions cultured in vitro, and indeed they could
not have been obtained in any other way. We know that these
responses involve not only T and B lymphocytes but also
macrophages, at the surface of which most antigens must be presented
if they are to evoke a positive response rather than to induce
tolerance. We also know that all three types of cell can elaborate
soluble factors which influence the behaviour or cause
differentation of each other. The relative importance and the
interplay of these regulatory factors is still far from clear;
in vitro studies can indicate what may happen but they are unable
to tell us what does happen in vivo. In real life, most antigens
become concentrated, and are largely broken down in macrophages,
and so far as the immune response is concerned, the macrophages
which matter are those located in lymphoid tissues. Anyone who
has made even a superficial study of lymph nodes or the spleen
will appreciate that these have a remarkably complex and delicate
architecture, which ensures that T and B lymphocytes encounter

macrophages and each other in restricted and probably special local environments, and that the flow of cells through them can be varied by changes in the channels through which they travel. The nature of the responses which result from such encounters is bound to be determined by these factors.

I am sure that the contributions which follow, describing the detailed anatomy and lymphocyte traffic through the spleen, will make clear with what a subtly and ingeniously constructed organ we are concerned. By way of illustrating the importance of micro anatomy for regulation of the immune response, I shall describe some recent work which indicates that germinal centres are sites especially effective for the generation of B-memory cells. (By these I mean long lived B cells whose production has been stimulated by a first encounter with an antigen, which do not secrete detectable amounts of antibody but when restimulated by a similar antigen, with T-cell help, can differentiate and secrete IgG antibody.)

Germinal centres were described in detail in 1885, by W. Fleming and his colleagues, who regarded these structures as sites of generation of leucocytes of the lymph, and clearly recognized that they varied with time in size and number.(1) Germinal centres are present in the spleen, lymph nodes and gut-associated lymphoid tissue of mammals, but their conservation during evolution implies that they fulfil some biologically important function.

Since B- and T-cell markers became available, several groups have confirmed that the lymphocytes in and immediately around germinal centres are B lymphocytes, although Weissmann demonstrated that a few T lymphocytes may be present in the periphery.(2) Germinal centres also contain cells with long finely branched cytoplasmic processes which extend between the lymphocytes. These "dendritic reticular cells" bind antigen-antibody complexes, but unlike macrophages they fail to ingest the complexes and retain them at their surface for weeks or even months. Several workers have speculated that the product of the dividing cells in germinal centres might be B-memory cells stimulated by antigens in the complexes retained on dendritic reticular cells. There is now strong, though still circumstantial, evidence that this is so.

The experiments which support this depend upon three findings. First, that attachment of antigen-antibody complexes to dendritic reticular cells depends upon activation of the third component of complement C3, and binding can be prevented by treating mice with cobra venom factor (CVF) to lower the C3 level below detectability;

second, that although CVF is a potent antigen, it is strictly thymus dependent so that in T-deprived mice (thymectomized, lethally irradiated and reconstituted with fetal liver cells) no anti-CVF is formed and consequently C3 levels can be kept down for weeks by repeated injection of CVF. Third, that T-deprived mice are fully able to generate B-memory cells in response to a primary antigenic stimulus, although they cannot make IgG antibodies. It was therefore possible to compare the generation of B-memory cells in T-deprived mice with normal C3 levels and in others without detectable C3. In the absence of C3 antigen-antibody complexes did not become localized in germinal centres - though the distribution of antigen as judged by radioautography was otherwise unchanged - and no B-memory cells were generated.(3) My colleague, Dr G.G.B. Klaus went on to show that preformed antigen-antibody complexes, at equivalence or in slight antigen excess, injected intravenously, were at least 100 times more effective at generating B-memory cells than was antigen alone, both in T-deprived and normal mice, provided that C3 was present. Furthermore, B-memory cells were generated several days earlier by preformed complexes than by antigen alone.(4) Our hypothesis is that in the primary response to an antigen there is a rapid T-independent production of sufficient antibody to generate circulating complexes with residual free antigen, part of which become localized via activated C3 on the dendritic reticular cells in germinal centres; the remainder being catabolized in macrophages elsewhere. The persisting antigen on dendritic reticular cells is able to stimulate virgin B cells, with receptors for the antigen, to divide and to generate B-memory cells. These have a higher density of receptors than their precursors and in general, the resulting B-memory cell population will have receptors with higher affinity for the antigen, since cells with such receptors will have been stimulated preferentially. The reason why B-memory cells stimulated by antigen localized in germinal centres do not go on to secrete IgG antibody is that this requires renewed contact with antigen and T-cell cooperation. Since T cells are effectively excluded from germinal centres, further differentiation will only take place if antigen and T cells are available in some other part of the lymphoid system or if the architecture of germinal centres is disrupted. Support for this hypothesis comes from experiments by J.G. Tew and his colleagues which have shown that labelled antigen injected into a foot pad persists for many months as antigen-antibody complexes on dendritic cells in the draining lymph node, long after detectable antibody production has ceased. However, if the draining lymph node is then removed, and fragments are cultured in vitro, substantial IgG antibody production begins after 2-3 days, without further addition of antigen.(5)

A further observation, which may have some general importance
for ideas about regulation of immune responses, has been made by
Klaus. He used a mouse myeloma protein M315, with reasonable
affinity for the hapten dinitrophenol (DNP), instead of anti-DNP
to make antigen-antibody complexes with dinitrophenylated
haemocyanin (DNP-KLH) as the antigen. When such complexes were
injected intravenously into normal mice not only were B cells
primed against DNP but also B cells reactive with the idiotypic
determinants of M315 protein. After administering complexes
containing a modest excess of M315 protein and subsequently boosting
with the latter alone, substantial amounts of anti-idiotype
appeared in the serum.(6) The system used is, of course, artificial
but since normal mouse antibodies against DNP would be a mixture of
unidentified idiotypes, the occurrence of anti-idiotypes in the
course of immunization against DNP-KLH would be undetectable.
What is interesting is the demonstration that antigen-antibody
complexes, once formed, may be expected to elicit antibodies
against the idiotypes of the antibodies which first appeared. This
could limit the antibody response to the origianl antigen DNP-KLH
in the manner predicted by Jerne's network theory.(7)

This introduction may appear to have strayed from the theme of
the role of the spleen, but I hope that I have made the point that
the microenvironment in which antigen, lymphocytes and other cells
(macrophages?) meet can determine the outcome. Germinal centres
are but one of the microenvironments in the spleen about which we
have some admittedly, superficial knowledge. The contributions
which follow will reveal others, at least as complex. To discover
their function will require a great deal of ingenuity, but perhaps
the changes which occur in parasitic infections will provide
important clues.

REFERENCES

1. Flemming, W. ET AL. "Studion uber Regeneration der Gewebe".
    Max Cohen & Sohn, Bonn (1885).

2. Weissmann, I.L. ET AL. Immune reactivity of lymphocytes. In:
    Advances in Experimental Medicine and Biology, (M. Feldman and
    A. Globerson eds.) Plenum Press, New York. 66: p. 229 (1976).

3. Klaus, G.G.B. and Humphrey, J.H. The role of C3 in the
    generation of B memory cells. Immunology, 33: 31 (1977).

4. Klaus, G.G.B. Generation of B memory cells with preformed antigen-antibody complexes. Immunology, 34: 643 (1978).

5. Tew, J.G. and Mandel, T. The maintenance and regulation of serum antibody levels: Evidence indicating a role for antigen retained in lymphoid follicles. Journal of Immunology, 120: 1063 (1978).

6. Klaus, G.G.B. Antigen-antibody complexes elicit anti-idiotypic antibodies to self-idiotypes. Nature, London, 272: 265 (1978).

7. Jerne, N.K. Towards a network theory of the immune system. Annales d'Immunologie, 125c: 373 (1974).

The Spleen

Leon P. Weiss, Department of Animal Biology, School
of Veterinary Medicine, University of Pennsylvania,
Philadelphia, Pa., USA

INTRODUCTION

The spleen is a haematopoietic vascular filter whose
distinctive stroma and vasculature regulate a large migratory
population of haematopoietic cells. The stroma acts as a
scaffolding which supports the vasculature and on which the
migratory cells are arranged. It is largely made up of branched
connective tissue cells, reticular cells, and associated
argyrophilic fibres, reticular fibres. The scaffolding or reticulum
formed by these cells and fibres is different in character in
different parts of the spleen. The parenchyma or pulp of the
spleen consists of the following major parts: the white pulp, which
is immunologically competent because it sequesters lymphocytes,
macrophages, and antigen and permits them to interact; the red pulp
which stores and tests red cells, granulocytes, and platelets; the
marginal zone which bears the heaviest vascular traffic or the pulp;
and ellipsoids, dense periarterial concentrations of phagocytes.
The spleen is enclosed in a heavy fibromuscular capsule from whose
internal surface branch a rich network of trabeculae. The
vasculature is the basis of organization of the lobule, the
structural and functional unit of the spleen. The lobule consists
of a lobular or central artery, its branches, an intermediate

---

Supported by United States Public Health Services, National
Institutes of Health Grant AM19920.

8

circulation, and an anastomosing system of draining veins, the
vascular or <u>venous sinuses</u>. The white pulp, the marginal zone,
the red pulp and the ellipsoids are organized upon this vasculature.

Please refer to the accompanying diagrams of the spleen in
reading the descriptions below.

WHITE PULP

The central artery enters the pulp from one of the many
trabeculae. As it runs through the pulp. it is surrounded by a

FIGURE 1

SCHEMATIC VIEW OF THE ORGANIZATION OF THE HUMAN SPLEEN

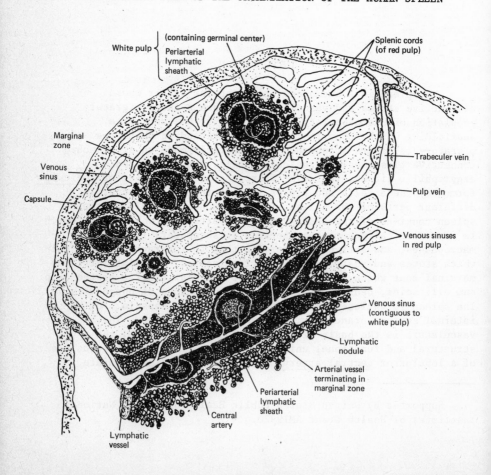

reticular sheath tightly packed with lymphocytes, the <u>periarterial</u> <u>lymphatic sheath</u>. This sheath is cone shaped, being broadest where the artery is largest, at its entrance into the pulp and gradually tapering to a point where is disappears, far out in the pulp where its artery is arteriolar in size. The lymphocytes comprising the periarterial lymphatic sheath emmigrate to the spleen from the thymus as disclosed both by thymic ablation and by labelling cells with an antithymic lymphocyte antibody. The periarterial lymphatic sheath, moreover, is deficient in lymphocytes in nude mice and in other thymic aplasias. Compact spherical collections of lymphocytes termed <u>lymphatic nodules</u> lie within the periarterial lymphatic sheath. These are rich in B lymphocytes as revealed by immunocytochemistry. It is within lymphatic nodules that germinal centres appear, thereby converting a primary to a secondary nodule. Germinal centres are sites of high-level antibody production

The white pulp has two components: periarterial lymphatic sheaths and lymphatic nodules. The latter may be made up of a germinal centre and a surrounding mantle zone. The white pulp is surrounded by the marginal zone. The remainder of the tissue depicted is the red pulp, which consists primarily of splenic sinuses separated by splenic cords. The pattern of blood flow is as follows. A trabecular artery enters the white pulp and becomes the central artery. The central artery passes through white pulp and gives rise to many branches. A few end within white pulp; some supply the germinal centre and mantle zone of the secondary nodule. Most terminate at the periphery of the white pulp, emptying in or near the marginal zone. A number of arterial vessels emerge from the white pulp, pass into the marginal zone, reach the red pulp, and curve back to empty into the marginal zone. Some arterial branches, in addition to the main stem of the central artery, run into the red pulp. Almost all terminate in the cords. Here, too, variation exists. Some arterial vessels terminate in a cord close against a sinus wall whereas others terminate in the midst of a cord, away from any sinus. Arterial vessels may terminate as capillaries or as somewhat larger vessels. Some arterial vessels may bear sheaths shortly before termination. The sinuses drain into pulp veins which, in turn, drain into trabecular veins. A sinus may abut the white pulp and receive lymphocytes or other free cells which migrate from white pulp across its wall and into its lumen. Efferent lymphatic vessels lie about the proximal portion of the central artery and run out of the spleen through the trabeculae. (From L. Weiss and M. Tavossoli, <u>Sem. in Hemat</u>., <u>7</u>: 372 (1970)

characteristic of a secondary response. The nodules and the
periarterial lymphatic sheaths, in aggregate, make up the white
pulp of the spleen.

ELLIPSOIDS

Most arterioles go on, past the periarterial lymphatic sheath,
reach capillary size and terminate. Before they terminate, they
may be ensheathed by a tight ellipsoidal nodule whose dominant
cell type is the macrophage. This sheath is prominent in pigs,
armadillos, dogs and cats. It is absent in mice and rabbits and
small and variably present in human beings. The term ellipsoid is
merely descriptive of the sheath's shape in certain species. In
fact, it can be spherical or of another shape, especially in specie
where it is not prominent. My colleague, Dr Julia Blue and I
suggest the substitution of the term periarterial macrophage sheath
for ellipsoid as functionally descriptive, applicable to its
appearance in any species, and parallel to the established and
useful term <u>periarterial</u> <u>lymphatic</u> <u>sheath</u>. Material passing throug
an arterial vessel within a periarterial macrophage sheath may flow
out through the vessel's luminal ending and be deposited in the red
pulp or the marginal zone. Or, as represented by such colloidal
or particulate markers as thorotrast or colloidal gold or even
such larger structures as intact red blood cells, the blood-borne
material may cross the endothelium before the termination of the
arterial vessel, enter the periarterial macrophage sheath and, if
it survives exposure to the macrophages of the sheath, pass through
to the surrounding pulp. While the macrophages of the periarterial
macrophage sheath represent an efficient collection of phagocytes
they are only one of the sets of macrophages occurring in the
spleen. As evident by the fact that these sheaths may be poorly
developed or absent in many species, their function may be assumed
by other parts of the spleen having concentrations of
reticuloendothelial cells. (See later discussion by Dr Weiss on the
reticuloendothelial system.)

The periarterial lymphatic sheath (PALS) in the centre of the
field is shown in cross section. Major branches of the central
artery (C art) are surrounded by a reticular meshwork consisting of
reticular cells and fibres. At the periphery of the pals the
reticulum takes on a circumferential character (crtl) in contrast t
the general branching character of the reticulum in the central
portions of the pals. Closely surrounding the central artery are
lymphatic vessels (ly). The central artery sends out branches
(art br).

FIGURE 2

SPLEEN - STROMA AND VASCULATURE

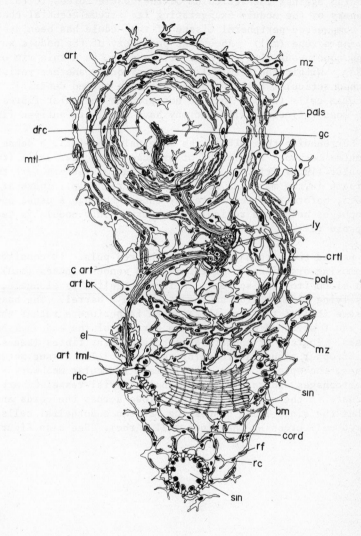

The stroma - consisting largely of reticular cells (rc) and reticular fibres (rf), and vasculature - of the major components of the spleen are shown in this diagram. The lymphocytes, macrophages, and other free cells (except for 2 red cells - rbc) have been omitted.

In this field a lymphatic nodule occurs within the pals. This nodule contains a germinal centre (gc). The germinal centre, expanding against the reticulum of the nodule forces it to the periphery of the nodule exaggerating its circumferential character. This compressed peripheral tissue of the nodule has been designated the mantle zone (mtl). About the periphery of the nodule surrounding it and continuous with the bulk of the pals, is a thin rim of pals (PALS'). Within the germinal centre may occur another variety of branched reticular cells. These have been termed dendritic reticular cells (drc). They do not produce reticular fibres (rf) as do most reticular cells and may have a role in antigen fixation.

Surrounding the PALS is the marginal zone (mz), a dense reticular meshwork fabricated of branched reticular cells (rc) and reticular fibres (rf). A great many arterial vessels (art tml) terminate in the marginal zone or in the red pulp. These arterial vessels, before they terminate may tunnel through a dense nodule of macrophages held in a reticular meshwork. This nodule is termed the ellipsoid (el).

Beyond the marginal zone lies the red pulp. It consists of an anatomosing organization of thin-walled venous sinuses (sin) which drain blood from the spleen. Their endothelium is elongate, the cells lying side by side as the staves in a barrel. The basement membrane (bm) of the vascular sinuses is perforate rather than a full sheet surrounding the outside surface of the endothelium of sinuses. In fact it appears as strong annular fibres (see surface view of upper sinus) surrounding the sinus. The tissue between the sinuses, the splenic cords (cord), is a reticular meshwork rich in macrophages and receiving terminal arterial vessels. Arteries terminate in the cords. The blood flows across the cords and then reaches the sinus lumen by passing between endothelial cells (see the red cell crossing the sinus wall - rbc). See also Figure 3.

13

FIGURE 3

RED PULP

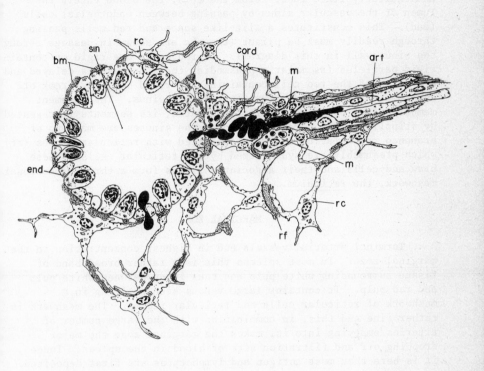

In this diagram a common type of intermediate circulation in
red pulp is shown. An arteriole (art) ends in the right (art term).
Between the arterial termination and the sinus (sin) on the left is
a portion of the splenic cord (cord) into which the arterial
termination opens. An efficient vascular pathway exists in the

cord between the ending of the arterial vessel and the sinus. It is bordered by the processes of reticular cells and by macrophages (m). Blood taking this pathway (see the red cells - rbc -) appear to flow through this segment of cord as efficiently as through an endothelially-lined tube. From the cord, the blood enters the lumen of the vascular sinus by passing between endothelial cells (end). This constitutes a slit-like space and red cells passing through readily must be pliant (two rbc are shown in passage through the sinus wall in this diagram). If red cells are rigid or contain rigid particles (as malaria plasmodia) their passage is delayed and a portion of the red cell containing the rigid body may break off and remain in the cords, just outside the sinus. The basement membrane (bm) of the sinus is fenestrated, its segments represented by stippling. The cords surrounding the sinuses are made up of branched reticular cells (rc) associated with reticular fibres (rf) which presumably are synthesized by the reticular cells. These branched cells and their associated fibres form a three dimensional meshwork, the reticulum.

MARGINAL ZONE

Terminal arterial vessels end in highest concentration in the marginal zone. In most spleens this is a rather broad band of tissue surrounding white pulp and thus lying between white pulp and red pulp. It contains large venous sinuses lying in a meshwork of reticular cells and reticular fibres. The meshwork is rather fine and this, in combination with the large number of arteries emptying into it, makes the marginal zone the major dropping off and filtration site of blood in the spleen. Indeed, it is here that most antigen and lymphocytes are first deposited.

RED PULP

The red pulp lies beyond the white pulp and marginal zone, in the farther part of the splenic lobule. It is typically the largest part of the spleen. The red pulp has three major components: 1) terminal arterial vessels; 2) an anastomosing system of vascular sinuses draining into major veins in the so-called sinusal spleens (represented by rabbit, dog, rat and the human being) or a scanty system of sinuses, in the non-sinusal spleens (exemplified by cat and mouse); and 3) a reticular meshwork lying between sinuses and veins, the splenic cords of sinusal spleens or the pulp spaces of nonsinusal spleens, into which arterial terminals discharge blood. The macrophage is typically the dominant migratory

cell in the cords or pulp spaces. Presumably blood monocytes
entering the cords through arterial terminations are selectively
held in the interstices of the reticular meshwork and there undergo
transformation into macrophages.

The typical pathway of an erythrocyte through the splenic
pulp is through the reticular meshwork of the marginal zone or red
pulp, into the lumen of a sinus or vein and then out. This course
constitutes a testing ground of red blood cells. The macrophages
loaded into the reticular neshwork are not only phagocytic, but
release hydrolytic enzymes and other materials into their
microenvironment. It has been suggested, however, that the oxygen
tension, cholesterol levels, and glucose concentrations of the red
pulp may become depressed and the pH and osmotic tension altered.
In some scanning electron micrographs, I have found crenulated red
cells, echinocytic according to the classification of Bessis,(1)
suggesting rough passage. Further "testing" of erythrocytes occurs
in their passage across the walls of the vascular sinuses. Here
the endothelial cells are elongate rods, reinforced by longitudinal
arrays of basal filaments, lying closely side by side and supported
by a markedly defective basement membrane. Dr Li-Tsun Chen and I
(2) have estimated that the slit between endothelial cells may be
widened to perhaps 0.5 micron in rabbits and rats for the transmural
passage of erythrocytes. Erythrocytes are normally pliant, but if
they contain rigid structures, as Heinz bodies or malarial plasmodia,
they may cross the sinus wall only with difficulty. Such an
erythrocyte may lie in the wall for a relatively long time, its
circulation through the spleen thus long delayed. It may break into
two fragments, the pliant portion going on into the sinus as a
spherocyte, and the portion containing the rigid structure remaining
in the perisinusal cord where it, and similar structures, accumulates
and may be phagocytized by cordal macrophages. One important
mechanism of splenomegaly is blockage of these interendothelial
slits with red cells containing rigid particles, with resulting
impedance to the circulation and congestive back-up in the cords.

While the sinus lumen is continuous with that of the major
draining veins (and, therefore, exit from the spleen), it is likely
that the lumen of the sinus is a holding place - a site of red cell
and platelet storage, a place for phagocytosis and cell maturation,
a likely site for increasing haematocrit and expressing fluid into
the surrounding pulp.

INTERMEDIATE CIRCULATION OF THE SPLEEN

The nature of the termination of arterial vessels, and the
nature of the circulation between the terminal arterial vessels

and the venous sinuses, the so-called intermediate circulation of
the spleen, has been a matter of intense study and controversy, the
matter of the so-called open vs. closed circulation. As
conventionally understood, in the closed circulation there is direct
connection between arterial termination and vascular sinus with
endothelial continuity. In the open circulation, on the other hand,
arterial endings open into pulp spaces or cords and then flow
through the wall of the vascular sinus or pulp vein, into its lumen.
Some variants have been developed. In the "divided circulation",
see Björkman (3), blood flows, presumably through closed pathways,
into venous sinuses. From the lumen of a sinus, fluid and small
particles flow out of the sinus through "stomata", or interendotheli
slits, in the sinus wall into the surrounding cords while such
larger structures as blood cells remain in the lumen. An
"anatomically open, physiologically closed" arrangement has been
surmised, see Weiss (4,5,6) and Snook (7,8) for review, wherein
flow from an open arterial ending, through cord and through
interendothelial slit in the sinus wall has the physiological
characteristics of flow in a closed system, despite the lack of
endothelial continuity, in an anatomically open system. See also
Fig. 3. Many studies of the intermediate circulation of the spleen
have depended upon injections of material into arteries or veins.
These studies of the vasculature include light microscopic studies
of latex, asphalt, or other coloured injection material to outline
blood vessels and the special case of injecting liquid plastic
(methacrylate), polymerizing it within vessels to obtain a high-
resolution cast and studying the cast by scanning electron
microscope after the tissue has been removed. In other experiments
of this sort, particles of different size have been injected
intravascularly into anaesthetized animals and, on the basis of
preferential distribution of particles, inferences on differential
paths of blood flow have been made. See references to Björkman, (3)
Knisely, (9) Snook, (7,8) Weiss, (4,5,6) and Chen (2,10) for further
discussion. The observations have, by and large, been sorted out,
although they are puzzling and apparently inconsistent. By light
or electron transmission microscopy of fixed, embedded, and
sectioned material, virtually all arterial endings are observed
to terminate in the red pulp or marginal zone as if the vessels
were sliced across (as the end of a garden hose), or with a flared
or ampullary ending - an "open" circulation. By the most fastidious
of vital microscopal techniques on the other hand, the arterial
vessels appear to connect directly with sinuses giving a "closed"
circulation. By means of vascular injection techniques and of
scanning electron microscopy interendothelial slits or apertures
have been recognized in the wall of vascular sinuses, vascular trave

through the pulp has been rationalized, establishing the existence of an "open" circulation and direct artery-venous sinus connections may also exist providing a "closed" component to the circulation.

I think the key to the interpretation of these findings on the so-called "open" circulation of the spleen is to recognize the cords and pulp spaces as vascular spaces and not as extravascular tissue. Investigations by electron microscopy have established that the vasculature of the body is diverse. Endothelium may be high, as in the post-capillary venules of lymph nodes and the arterial system of the spleen, or attenuated and perforate as in the renal glomerulus. The subcapsular lymphatic sinus in lymph nodes is criss-crossed by reticular cells and fibres. The cavernous vessels of the genital system are broad trabeculated spaces. The venous vasculature of the ear lobes is patulous. It is consistent with the modern view of the vasculature to define the splenic cords and pulp spaces as vascular spaces. It is clear from a number of lines of evidence, moreover, that flow through a cordal space can be as efficient and even as cylindrical as through a conventional tubular vessel. The technique of vital microscopy, moreover, a low-resolution method, may not be able to distinguish a column of blood flowing through a cord and through interendothelial slits of the wall of a venous sinus from that flowing in conventional vessels. I believe, therefore, that the matter of an open or closed circulation in the spleen is, in essence, unreal. There is no extravascular or open flow. The splenic cords are vascular passages. It remains to define in a spleen the relative contributions the two components of the intermediate circulation – the component carrying blood across the pulp spaces and the component confining blood to endothelial-lined tubes – make to the flow of blood in given species, under given circumstances.

LYMPHATIC VESSELS

The problem of the deep lymphatic pathway in the spleen has been with us for more than 250 years, long predating the dilemma regarding the continuity or discontinuity of blood circulation. Although there is no shortage of evidence as to the existence of deep lymphatic vessels in the spleen, most textbooks still reflect the view that splenic lymphatics can be found only in the immediate vicinity of the hilus, and even then in only a few mammals, or that they cannot be found in the splenic pulp at all. And yet, Barcroft and Florey, as early as 1928 offered evidence of lymph being formed in the spleen and Halta et al. succeeded in isolating lymph from the spleen and analyzing it. My colleague, Vaclav Janout of Charles University and I (11) have demonstrated lymphatic vessels penetrating deep into the white pulp along arterial vessels in the spleen of the marmot by light and electron microscopy, have shown these efferent vessels may contain large numbers of lymphocytes, and that lymphocytes appear active in crossing the lymphatic wall. Snook (7) has described and figured deep lymphatics in a variety of animals. Dr Norman Anderson and I (paper in press) have shown that these lymphatic vessels are indeed an efferent pathway for lymphocytes by labelling thoracic duct lymphocytes in the rat, injecting and charting their pathway through the spleen by autoradiography. A large component of lymphocytes quickly exits via the venous system. Another substantial component, over a period of 3 to 4 hours, has a more selective pathway. These lymphocytes move to the marginal zone, cross the white pulp and exit via lymphatics. As they cross the white pulp, T and B cells sort out into T and B cell zones. This lymphocytic pathway in spleen is not unlike that in lymph nodes: in each case, from blood to lymph, separation into T and B cell zones, and the passage of about 4 hours for T cell movement and a longer time for B cells.

Antigen is channelled distinctively in the spleen as well. Ten minutes after injection in a secondary response and one hour in a primary. Nossal and his colleagues found polymerized flagellin concentrated in the marginal zone. Within seven days in a secondary response it moved into the white pulp and capped the germinal centre. See Weiss (12) for further discussion.

STROMA

The term reticular cells, as used here, refers to the major cell type forming the stroma of the spleen and other haematopoietic tissues. This cell type is evidently fibroblastic, as indicated by extracellular fibrillar matrix (reticular fibres) lying upon its surface, and by its content of rough ER whose cisternae often contain a granular material resembling this extracellular matrix.(13) Reticular cells have been considered phagocytic. But their uptake of colloidal material and of damaged erythroid cells is decidedly limited. In contrast to macrophages they have few lysosomes. Other dendritic cells may be recognized in haematopoietic tissues. Steinman et al. (14) have identified a novel dendritic cell present in small numbers in the white pulp of the spleen. It is non-fibroblastic, non-phagocytic, and may be isolated, in part by its capacity to adhere to glass. It is identified in electron micrographs of intact white pulp by subtle morphological criteria. It may have a role in fixing antigen. Thus, reticular cells as we have defined them in this paper, like fibroblasts, osteoblasts, fat

19

cells, and smooth muscle cells, and certain epithelia appear to elaborate extracellular fibres. This capacity and their formation of a spongework accounts for their role in providing a scaffolding for blood, haematopoietic and connective tissue cells in spleen, lymph nodes, thymus and bone marrow.

But fibroblastic cells have major functions in addition to supporting parenchyma and producing fibres with tensile properties. They possess, during development an enormous capacity for induction. Local fibroblasts induce the histotypic conformation of many epithelia, as in salivary gland, renal tubules, and teeth. (15) From the tissue culture and transplantation studies of Friedenstein et al. (16) and from the electron microscopic work on the stroma of the bone marrow of my laboratory, it would appear that fibroblastic cells of the marrow, likely reticular cells, are necessary to induce cellular differentiation (17) and to sequester stem cells from the circulation.

It is evident that there are diverse pathways through the spleen. Migratory cells (and particulate material) are rigorously compartmentalized and so set up that they selectively interact and differentiate. The distinctive microenvironment of the spleen that permits these phenomena of homing, cell sorting, cell interactions and cell differentiations are operative and may well be in large part due to the reticular cells forming its distinctive reticular meshworks. It is these meshworks through which the migratory cells move, in which they are compartmentalized, interact and differentiate.

REFERENCES

1. BESSIS, M. Red cell shapes: An illustrated classification and its rationale in red cell shape, physiology, pathology, ultrastructure. (Bessis, M., Weed, R., and Leblond, P.F. eds.) Springer-Verlag, Berlin (1973).

2. CHEN, L.T. and WEISS, L. Electron microscopy of red pulp of human spleen. American Journal of Anatomy, 134: 425 (1977).

3. BJÖRKMAN, S.E. The splenic circulation. With special reference to the function of the spleen sinus wall. Acta Medica Scandinavica (Suppl. 191) 128 (1947).

4. WEISS, L. Structure of Intermediate Vascular Pathways in the Spleen of Rabbits. American Journal of Anatomy, 113: 51 (1963).

20

5. WEISS, L. A scanning electron microscopic study of the spleen. Blood, 43: 665 (1974).

6. WEISS, L. The Spleen. In: Histology, 4th Edition (Weiss, L. and Greep, O.R. eds.) p. 545 (1977).

7. SNOOK, T.A. A comparative study of the vascular arrangements in mammalian spleens. American Journal of Anatomy, 87: 31 (1950).

8. SNOOK, T. The histology of vascular terminations in the rabbit spleen. Anatomical Record, 130: 711 (1958).

9. KNISELY, M.H. Spleen Studies. I. Microscopic observations of the circulating system of living unstimulated mammalian spleen. Anatomical Record, 65: 23 (1936).

10. CHEN, L.T. Microcirculation of the spleen: An open or closed circulation? Science, 201: 157 (1978).

11. JANOUT, V. and WEISS, L. Deep splenic lymphatics in the marmot: An electron microscopic study. Anatomical Record, 172: 197 (1972).

12. WEISS, L. The cells and tissues of the immune system. Prentice-Hall, Inc., Englewood Cliffs, N.J. (1972).

13. PICTET, R. ET AL. An electron microscopic study of the perfused fixed spleen. I. The splenic circulation and the RES concept. Zeitschrift fuer Zellforschung und Mikroskopische Anatomie, 96: 372 (1969).

14. STEINMAN, R.M. ET AL. Identification of a novel cell type in peripheral lymphoid organs of mice. IV. Identification and distribution in mouse spleen. Journal of Experimental Medicine, 141: 804 (1975).

15. GROBSTEIN, C. Tissue interaction in the morphogenesis of mouse embryonic rudiments in vitro. In: Aspects of Synthesis and Order in Growth. Princeton University Press. p. 233 (1954).

16. FRIEDENSTEIN, A.M. ET AL. Stromal cells responsible for transferring the microenvironment of the hematopoietic tissues. Transplantation, 17: 331 (1974).

17. TRENTIN, J.J. Influence of hematopoietic organ stroma (hematopoietic inductive microenvironments) on stem cell differentiation. In: Regulation of Hematopoiesis, Vol. 1 (A.S. Gorden ed.) Appelton-Century-Crofts, N.Y. p. 161 (1970).

The Reticuloendothelial System

Leon P. Weiss, Department of Animal Biology, School
of Veterinary Medicine, University of Pennsylvania,
Philadelphia, Pa., USA.

The concept of the reticuloendothelial system, as developed by
Aschoff, Kyono and others, is that of an extensive system of
protection and regulation whose salient feature is phagocytosis.
They expressed this system in terms of three major cell types: the
macrophage, the stromal reticular cell and the endothelial or lining
cells of the vascular sinuses in the spleen, bone marrow, lymph
nodes and other haematopoietic tissues. It may be useful to present
the ideas of the role of these cells which underlay the development
of the concept of the reticuloendothelial system and to compose
them to current knowledge and speculation.

Macrophages were identified, as established by Metchnikoff
as large-capacity phagocytes different from the granulocytes or
microphages. The macrophage was recognized as a wandering cell and,
in the exuberence of the early days, this cell was defined in terms
of the technique used and, for a long time, cells were recognized as
different which were, in fact, similar. Thus we had the pyrrhol
blue cell - which took up pyrrhol blue; the wandering cell - because
it moved about; and the resting-wandering cell, because the wandering
cell would stay still for a while. There was the cell, the
closmastocyte, the adventitial cell, the histiocyte, recognizing such
characteristics as closeness to blood vessels, the tendency to
break off bits of cytoplasm during movement, the connective tissue
nature and the fact that, though a phagocytic cell, it could be

found without having engulfed material. Maximow recognized the monocyte as a source of macrophage, which he delineated as polyblasts. Hofbauer described a vasculated cell in the placenta, the Hofbauer cell. As the dust has settled we recognize all of these cells as macrophages.

But our concept of macrophage is broadening to encompass far more than phagocytosis. Macrophages are secretory cells. John Humphrey and I commented on this at the beginning of the meeting. Macrophages synthesize and secrete collagenase and probably hyaluronidase, components of complement, interferon and other materials. Indeed lysosomes, used to digest phagocytozed substances, can be viewed as secreting granula which on secretion provide a bath of acid hydrolytic enzymes in the microenvironment of macrophages. This is notable in the red pulp of the spleen. Other non-phagocytic functions include their facilitation of T and B cells interacting in immune responses. Macrophages, moreover, as discussed earlier, will probably be found to be divisible into subsets, as have T and B cells, fibroblasts, etc. One clear subset about which there is a great deal of information is the alveolar macrophage of the lung - capable of heightened oxidative metabolism.

Not only are macrophages more than phagocytic cells they have, as is the case with many of the parasites considered at this meeting, a complex life cycle. They begin as promonocytes in the marrow, circulate as monocytes, readily undergo transformation into macrophages - but they can go on to form epithelioid cells and multinucleate giant cells. Such multinucleate giant cells include the Langhans giant cell and, also, the osteoclast. These cells are often not phagocytic. They lose lysosomes, accumulate increased plasma membrane and many mitochondria. They turn from phagocytic cells to cells specialized in active transport, the transport of Ca^{++} for example, and in extracellular digestion, such as heightened collagenase activity in the case of osteoclasts.

To turn to the reticular cells - the dendritic cells of the haematopoietic system. The early views of these cells were that, in forming a reticulum in association with reticular fibres, a scaffolding supporting the free cells was provided. But these reticular cells were also considered to be the haematopoietic stem cell - to round off the reticulum and transform into any of the blood cells - this notion has been disproved. These reticular cells were also regarded as phagocytic - the so-called fixed macrophages. It is now increasingly evident that these cells are in fact not phagocytic - so which is there left for the reticular cell? I think a great deal. It remains the scaffolding cell. It forms a mehurst.

It probably compartmentalizes the migrating cells and directs their movement - the idea of homing. It probably induces cell differentiation in the haematogenetic system. Again, as with macrophages, we shall probably find different kinds of dendritic cells. Steinman's work on the non-phagocytic, non-fibre producing, Ag binding dendritic cell of splenic white pulp has been referred to earlier.

What of the endothelium? The earlier workers thought the endothelium lining the vascular sinuses in haematopoietic tissue so special they considered the common term "endothelium" inappropriate and brought forth such terms as "lining cell" and "littoral cell". They considered that this endothelium was phagocytic. In fact it is not. But these vessels do have special functions. They permit the selective movement of cells across the vascular wall, as in the bone marrow where only mature cells are commonly permitted into the vascular lumen. They test red cells as in the spleen and "pit" malaria plasmodia.

There are thus no "fixed" macrophages in that the reticular cell and endothelial cells of the sinuses are not phagocytic. Even the Kupffer cell is a transient cell, infiltrating the endothelium of liver sinusoids for a limited time, coming from bone marrow and travelling through the blood.

There have been efforts to rename or reclassify the RE system. The term "mononuclear phagocytic system" has been proposed. I consider this inappropriate since "mononuclear" means, simply, one nucleus - although it is used to mean a round or near round nucleus. In fact granulocytes and megakaryocytes are mononuclear cells. The "phagocytic" part of this term is insufficient to cover the many other functions of this system. The term "reticuloendothelial system" is, in many respects, unsuitable. It fails to recognize one of the protagonists, the macrophage. But it is, nonetheless, a serviceable term. There is, to my mind, no better term at this time and we are, I hope, about to learn a good deal more about this system. After all, including the sessile stroma and vascultine of the haematopoietic system as it does, it must be central to control of and delivery of T and B cells and other elements in haematopoiesis. All of the techniques used for elucidation of the migratory cells, for characterizing lymphokines and haematopoietic humoral substances as erythropoietin can be applied to these sessile elements. I recommend we stay with the term reticuloendothelial system at this time and, let's hope, in a few years, we shall be able to make a definitive characterization of this system and introduce an improved name.

DISCUSSION

HUMPHREY: I would like to describe what appears to be yet another specialized anatomical environment in the spleen, to supplement Professor Weiss's beautifully clear description of the functional anatomy of the spleen.

In mice and rats, at least, the macrophages of the marginal zone of the spleen white pulp are functionally and morphologically distinct from those in the red pulp. The evidence for this came in the course of studies (J.H. Humphrey, to be published) to test the ability of conjugates of the hapten dinitrophenyl (DNP) with a variety of polysaccharide carriers to inhibit secondary anti-DNP antibody responses to DNP protein conjugates. This was an extension of the observation that DNP conjugates of the pneumococcal capsular polysaccharide SIII are particularly effective at inducing unresponsiveness to DNP even in already primed mice.(1) The mechanism whereby polymeric hapten conjugates of thymus-independent carrier molecules can render unresponsive B-cells with receptors for the hapten has been reviewed elsewhere.(e.g. 2) Conjugates were prepared of DNP with several uncharged and negatively charged (anionic) polysaccharides, and these were also trace labelled with ^{131}I or ^{125}I via added tyramine residues in order to study their metabolism and detailed distribution in different tissues. It became apparent that DNP conjugates of negatively-charged polysaccharides were generally quite effective inhibitors of secondary IgG anti-DNP responses, whereas similar conjugates of neutral polysaccharides tended to elicit prolonged synthesis of small amounts of IgM anti-DNP antibodies, and were only tolerogenic in normal mice when relatively large amounts were administered. All the conjugates were retained undegraded in the body with long half lives (20-60 days), and all disappeared rapidly but not quite completely from the circulation. However, the distribution of the charged and uncharged conjugates differed markedly. As judged by autoradiography, the acidic polysaccharides were localized in what are conventionally regarded as macrophages, whereas the neutral polysaccharides were notably absent from liver Kupffer cells, and strikingly concentrated in macrophages of the marginal zone of the spleen white pulp and the marginal sinuses of lymph nodes but not in spleen red pulp macrophages or those of lymph node medullary cords. In order to examine these differences further, fluorescein and rhodamine conjugates were prepared of SIII and of two neutral polysaccharides Ficoll and hydroxyethyl starch, at a low substitution ratio (about 2 residues/100,000 daltons) so as to avoid altering their properties. After intravenous administration these showed similar distributions to their radiolabelled

counterparts, but had the advantage that cells containing them could
be examined by fluorescence microscopy in frozen sections of tissues
or in unfixed cell suspensions. Because the conjugates were only
very slowly degraded intracellularly, the cells which had taken them
up could be identified days or even weeks after the initial
administration. A preliminary account of these experiments in in
press.(3) For the purpose of this discussion, the main point is
that the fluorescent Ficoll or starch became localized in the spleen
exclusively in large branching cells in the marginal zone of the
white pulp, in contrast to SIII which was mainly concentrated in red
pulp macrophages. When single cell suspensions are made by teasing
the spleen apart in the usual manner, marginal zone macrophages are
mostly either broken - presumably because of their branching
structure - or are left behind as sheets of cells in the debris.
Only by perfusing with collagenase has it been possible to obtain
these cells single and intact. They make up about one tenth of the
total macrophage population, and are generally much larger than other
spleen macrophages, with a large oval or circular nucleus and
extensive cytoplasm which stains strongly positive for acid
phosphatase and non-specific esterase. A striking feature was that
lymphocytes tended to adhere around the marginal zone macrophages
both in suspension and after attachment to glass. The lymphocytes
formed Fc and EAC rosettes and were presumably B-cells. Since the
marginal zone is recognized as an area of B-cell traffic, this
adherence may be significant for their re-routing into the B-cell
areas of the white pulp.

MITCHELL: Mr Chairman, does the extraordinary plasticity of
uninfected erythrocytes referred to by Professor Weiss have
implications for the separation of merozoites from erythrocytes by
nucleopane filtration techniques?

COHEN: Mammal red cells are deformable, whereas parasitized
erythrocytes of some malarial species are relatively rigid.
Parasitized blood can be depleted of normal red cells by passage
through a 2μ filter; merozoites liberated subsequently can be
harvested by the same sieving technique.

MILLER: I was very interested by the heterogenesis of macrophages
which both Professor Weiss and Professor Humphrey discussed. You
will recall the very nice work of Rosenthal and collaborators on
antigen-induced proliferation of sensitized T cells in vitro by
antigen-released macrophages. These macrophages were shown to have
receptors for lymphocytes and actually formed rosettes with
lymphocytes. Is it known whether marginal zone or ellipsoid
macrophages differ from red pulp macrophages in their capacity to

form rosettes with lymphocytes? Furthermore, since I-associated (Ia) antigens appear essential in the activation of T lymphocytes, is it known whether the distribution of Ia components is different in the two types of macrophages?

WEISS: I do not believe it is known if marginal zone macrophages or ellipsoidal macrophages form rosettes with lymphocytes. I suspect the former may and the latter may not, simply because lymphocytes regularly occur among the marginal zone macrophages and not among those in ellipsoids. In some cases, however, specifically ellipsoids in the chicken as has been shown by R. White, the ellipsoid is carried on an arterial vessel so recurrent that it presses against the white pulp and, in fact, the white pulp may actually partially surround it. These ellipsoidal macrophages are well set up to interact with lymphocytes, in contrast to those ellipsoids out in red pulp. I know of no information on the distribution of Ia associated components on the surface of macrophages.

BATTISTO: Since the reticulum cell in the spleen is in such close association with T cells, as you have so elegantly shown, do the two cells communicate? Have you any evidence that information is being passed by the two sorts of cells?

WEISS: It is, I suspect, quite possible that T cells and reticular cells communicate, possibly by means of the sialic acid or proteoglycons in the cell coats. These two cells, moreover, are the only cells present in many sites of T-cell sorting and T-cell migration. There are, however, no data on any information passed between these cell types.

BATTISTO: Professor Humphrey, can you explain why hapten-conjugated acidic polysaccharides are more effective at suppressing anti-hapten responses than are the hapten-conjugated neutral polysaccharides.

HUMPHREY: The mode of action of hapten-conjugated SIII has been studied in detail by Klane and Abbas and appears to involve a time- and concentration-dependent inhibition of regeneration of B-lymphocyte specific receptors for the hapten, in the absence of T-cell help (since SIII is unable to generate T-cell help). The mechanism is the same as that involved in "clonal abortion" and effector cell blockade, described by Nossal and his colleagues. The neutral polysaccharides may act in the same way, but the concentration must be higher (at least where adult B cells are involved). It may be that the limited localization of the neutral polysaccharides implies much more limited contact of B cells with them, but this is only a guess.

ALLISON: Professor Humphrey, in the data on mice that you presented immune complexes persist for long periods. Could this be relevant to the persistent stimulation of B lymphocytes in the TSS? Also, Gerry Klaus found that immune complexes can, in the absence of adjuvant, stimulate anti-idiotypic responses, some with broader specificity than the antigen-combining site. Is it possible that the complexes in TSS likewise stimulate anti-idiotypic responses with relatively broad specificity that can function as polyclonal mitogens?

HUMPHREY: The complexes which persist and stimulate B-memory cell production are those in germinal centres. We did not have any ·evidence that they stimulated pre-existing B-memory cells - i.e. kept a response ongoing - although we equally do not show that they do not do so. As I pointed out in my introduction, the B-memory cells require further contact with antigen and T-cell cooperation before they differentiate into antibody producers. In such a complex situation it is difficult to give a straight yes or no answer to your first question. Perhaps the first thing to do would be to see whether malarial antigen-antibody complexes are detectable in germinal centres in the TSS. As regards your second question, if anti-idiotype antibodies were produced, and if human anti-idiotype can stimulate T or B cells with the right idiotype to proliferate, then the answer would be that your suggested mechanism is possible. But there are two "ifs", and I do not know whether either actually occurs.

REFERENCES

1. MITCHELL, G.F. ET AL. Inhibition of secondary anti-hapten responses with the hapten conjugated to type III pneumococcal polysaccharide. European Journal of Immunology, 2: 460 (1972).

2. KLAUS, G.G.B. B cell maturation: its relation to immune induction and tolerance. in: B and T cells in Immune Recognition (F. Lorr and G.E. Roelants, eds.) John Wiley & Sons Ltd., Chichester (1977).

3. HUMPHREY, J.H. Marginal zone and marginal sinus macrophages in the mouse are distinct populations. in: Proceedings of the VIth International Conference on Lymphatic Tissues and Germinal Centres in Immune Reactions (Muller Ruchholtz, W. and Muller-Hermelink, H.K., eds.) (1978).

Lymphocyte Recirculation Between the Spleen and the Blood

W.L. Ford and M.E. Smith, Department of Pathology,
University of Manchester, Manchester, England.

INTRODUCTION

Among the major sites of lymphocyte recirculation the spleen
is predominant. The other organs within which lymphocytes
recirculate from blood to tissue and back to blood on a large
scale are the lymph nodes, gut-associated lymphoid tissue and the
bone marrow. However in the adult rat the flux of lymphocytes
recirculating through the spleen is greater than the fluxes through
all the other compartments combined.(1)

The most direct evidence for the high flux of lymphocytes
between the blood and the spleen comes from the type of experiment
pioneered by Gowans and Knight (2) in which thoracic duct
lymphocytes (TDL) are radiolabelled in vitro and returned to a
syngeneic recipient by intravenous injection. When ^{3}H-uridine is
used to label TDL in vitro most of the isotope is taken up by
T lymphocytes. At 1-2½ hours after injection 50-55% of the
injected radioactivity can be recovered from the spleen of a
recipient.(3) This high proportion of radioactivity in the spleen
falls by more than half by 12 hours after injection and by three-

---

Abbreviations used: TDL - Thoracic duct lymphocyte
                      PALS - Periarteriolar lymphoid sheath
                      MZ - Marginal zone

quarters by 24 hours after injection. Only a small amount of the fall in radioactivity between 2½ and 24 hours can be accounted for by loss of isotope from the cells. By contrast the radioactivity recoverable from lymph nodes steadily increases throughout this period. (3)

Bradfield and Born (4) used a number of experimental approaches to establish that lymphocytes were physiologically redistributed from spleen to lymph nodes beginning a few hours after injection and that no other explanation for the steep fall in radioactivity recoverable from the spleen was plausible. The early decrease in splenic radioactivity suggests a transit time of T lymphocytes migrating through the splenic pulp of only a few hours. The data from these experiments in vivo are consistent with the estimated modal transit time of approximately five hours originally based on results obtained with the isolated, perfused rat spleen. (5) Subsequently, analogue computer analysis of these results improved the quantitative conclusions which could be drawn regarding the kinetics of lymphocyte exchange between the isolated spleen and the perfusate. (6) Another approach to quantifying the recirculation of lymphocytes was by confining a field of β-irradiation to the spleen. (7) This technique confirmed that the flux of lymphocytes through the spleen was greater than the thoracic duct output and also suggested that at least 95% of small lymphocyte in the thoracic duct population were capable of migrating through the spleen.

The dramatic reduction in the number of T cells in the spleen between three and 12 hours after intravenous injection is obvious on inspection of autoradiographs. However when B lymphocytes were separated from TDL and labelled in vitro with ^{3}H-leucine there was no perceptible reduction in the number of cells in the spleen for up to 24 hours after injection. This suggests that B lymphocytes take much longer to traverse the splenic pulp and return to the blood than do T lymphocytes (8) but more precise evidence on this point is lacking.

THE ROUTES OF B AND T LYMPHOCYTE MIGRATION

Several workers have agreed that the initial localization of B and T lymphocytes in the spleen is indistinguishable. Within 10-15 minutes of intravenous injection most labelled lymphocytes are present in the marginal zone (MZ). (8,9,10) T cells enter periarteriolar sheaths (PALS) rather quickly so that by 30 minutes after injection most T cells are in that compartment. The smallest PALS and the periphery of the larger PALS receive T cells before the

centre of the large PALS suggesting that the predominant direction
of migration is from the periphery to the hilum, that is opposite
to the direction of arterial blood flow. B lymphocytes also enter
the periphery of the PALS and remain closely intermingled with
recirculating T cells for several hours after entering the spleen.
Then B cells accumulate at the junctional region between the PALS
and the lymphocyte corona (or mantle zone) around germinal
centres. At about 3-6 hours after entering the spleen B lymphocytes
segregate from T cells by entering the lymphocyte corona.(8) The
germinal centre and the enveloping corona is a polarized structure
and the last area to be reached by migratory B lymphocytes is the
distal part of the corona. At 24 hours after injection B
lymphocytes are seen to be distributed rather evenly around the
lymphocyte corona. A small minority are in the middle of the
germinal centre but whether they have reached there after following
the same route as the majority of B cells or have entered directly
by crossing the walls of small blood vessels within the germinal
centre is still to be decided.

THE ROUTE OF RETURN OF B AND T LYMPHOCYTES
FROM THE SPLEEN TO THE BLOOD

Recirculating lymphocytes returning to the blood from lymph
nodes are believed to do so exclusively via efferent lymphatics
which contribute to major lymph conduits such as the thoracic duct.
(11) Lymphocytes recirculating through gut-associated lymphoid
tissues and through non-lymphoid tissues such as the skin are also
assumed to return via long lymphatic channels. By contrast,
lymphocytes, returning to the blood from the extravascular spaces
of the spleen and bone marrow are believed to do so within the
confines of each organ. In each case, the highways taken by
lymphocytes probably involve channels which would be regarded as
lymphatic passages by histological criteria.

Two independent groups have studied the isolated, perfused
spleen of the pig (12) and the rat.(5,13) In both cases very large
numbers of lymphocytes were released by the spleen into the venous
effluent although the lymphatic channels in the vascular pedicle
of the spleen were ignored and presumably obstructed. Nevertheless
it cannot be excluded that many lymphocytes which would have taken
a lymphatic route in an intact animal leave the perfused spleen by
a more direct route.(14)

The fine lymphatics in the vascular pedicle of the spleen have
been thought to drain only the capsule and trabeculae of the spleen
but Weiss (14) has suggested that a substantial number of lymphocytes

returning to the blood, perhaps a third of those leaving the spleen, take this route and eventually reach the thoracic duct. The strongest objection to this view is based on the temporal pattern of recovery of lymphocytes from a thoracic duct fistula after intravenous injection. The peak recovery in splenectomized recipients is at 15-18 hours after the injection of labelled lymphocytes;(15) very few are recovered in the first six hours (Fig. 1). If lymphocytes leaving the spleen were contributory to the thoracic duct output then, in an intact recipient, one would expect to recover many cells at around 4-6 hours after injection, when it is generally agreed that most T cells are leaving the spleen. Moreover the time of maximum recovery would be expected to occur earlier in the intact compared to the splenectomized recipient.

The temporal pattern of labelled lymphocyte recovery from a thoracic duct fistula has been compared in splenectomized and intact rat recipients by W.L. Ford and also by S.E. Dorsch and B.J. Roser (both unpublished). They found that the presence of a spleen only slightly altered the pattern of recovery but in the opposite way from that which would would predict if lymphocytes returned via the thoracic duct. (Fig. 1). The spleen damps down the high blood levels early after injection and thus delays the migration of lymphocytes through lymph nodes and non-lymphoid tissues into the thoracic duct. A similar line of argument is that in the pig the flux of lymphocytes between the blood and the spleen was estimated to be 67 times the thoracic duct output.(12) Although the structure of lymph nodes and their tiny contribution of cells to the thoracic duct is anomalous in this species, the structure of the spleen and its capacity to support a recirculation of lymphocytes does not seem to be exceptional.

In the mouse spleen, bridging channels across the marginal zone were described by Mitchell.(15) She found that labelled lymphocytes were concentrated there 4-8 hours after intravenous injection, suggesting that these channels conveyed recirculating lymphocytes from the pseudolymphatic spaces of the PALS back across the MZ to the venous sinuses of the red pulp. This work provides a reliable indication of the route of exit of T cells from the white pulp but there is still no clue about the route by which B lymphocytes leave the lymphocyte corona.

RECIRCULATING AND NON-RECIRCULATING LYMPHOCYTES IN THE SPLEEN

After prolonged thoracic duct drainage to deplete the recirculating pool the spleen does not lose all of its lymphocytic

FIGURE 1

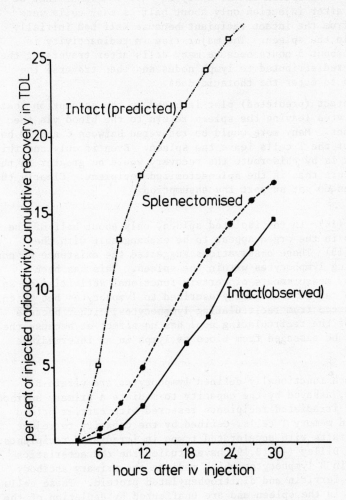

Thoracic duct lymphocytes were labelled _in vitro_ with ^{3}H-
uridine and injected intravenously into splenectomized or intact
syngeneic recipients which themselves had been subjected to thoracic
duct cannulation. The mean cumulative recovery of cell-associated
radioactivity in lymph is plotted over a period of 30 hours. If

the splenectomized recipient is regarded as the standard the influence of the spleen can be analysed. It was found that at 6-12 hours after injection only about half as many cells were recovered from the intact recipient because half had initially localized in the spleen. The major rise in radioactivity is delayed by about 5 hours because many cells after traversing the spleen are redistributed to lymph nodes and then traverse the lymph nodes to enter the thoracic duct.

The intact (predicted) plot is based on the assumption that the lymphocytes leaving the spleen return to the blood via the thoracic duct. Many more would be recovered between 6 and 12 hours when most of the T cells leave the spleen. Even if only one third of the flux is by this route the recovery would be greater in the first 12 hours than in the splenectomized recipient. Clearly the observations do not support the assumption.

population.(16)   In the isolated spleen, only about half of the lymphocytes in the organ appear to be exchangeable with the perfusate.(13)   These observations suggested the existence of non-recirculating lymphocytes within the spleen. This has been confirmed by measurements of certain functional activities of spleen cells which can be confidently ascribed to lymphocytes but which can be divorced from recirculating lymphocytes, either because depletion of the recirculating pool has no effect or because the cell cannot be passaged from blood to lymph in an intermediate recipient.

Two such functionally defined lymphocytes are virgin B lymphocytes, assayed by the capacity to mediate a primary antibody response in irradiated recipients restored with carrier primed T cells, and memory T cells, defined by the capacity to reject heart allografts with accelerated tempo in irradiated recipients. Strober and Dilley (17,18,19) have studied the characteristics of the virgin B lymphocytes which mediate the primary antibody response to ferritin and dinitrophenylated protein. These cells are present in the spleen and are unaffected by depletion of the recirculating pool. They cannot be passaged from blood to lymph in an intermediate rat although they are present in thoracic duct lymph.

More recently Roser and his colleagues (20,21) have applied similar techniques to study the T cells in alloimmunized rats which, after transfer in small numbers to irradiated recipients, provoke prompt rejection of heart allografts. These cells were found in the spleen and thoracic duct population but failed to

recirculate from blood to lymph. Incidentally, they were "long-
lived" as assessed by treatment of the immune donor with
vinblastine in contrast to the rapidly turning over virgin B
lymphocytes.

Not all T cells generated as a consequence of immunization have
the same characteristics. Helper T cells in rats primed to a
number of conventional antigens readily recirculate from blood to
lymph; in fact they may be enriched by this procedure.(17) Because
the term "memory T cell" has led to confusion in the past, it may
be suggested that the memory cells in Roser's system of allograft
rejection may turn out to belong to the $T_{c/s}$ subset (22) and may
correspond to the immediate precursor of a cytolytic cell that
has been demonstrated in immunized mice.(23) On the other hand,
the T memory cells contributing to secondary antibody responses
probably belong to the T_H subset. Both the T_H and $T_{c/s}$ subsets
are present in the spleen but they may turn out to have strikingly
different migratory properties.

THE SELECTIVE ENTRY OF RECIRCULATING LYMPHOCYTES INTO THE SPLEEN

Both the spleen and lymph nodes have efficient mechanisms for
distinguishing recirculating lymphocytes from the other cells in
the blood so that it is only lymphocytes which are selected to
enter tissues while passing through the blood vessels. The
specialized vascular endothelia responsible for sorting out
leucocytes are morphologically very different in the spleen and
lymph nodes. In lymph nodes lymphocytes migrate across a plump,
high walled endothelium in post-capillary venules within the thymus
dependent areas.(2,24) In the spleen lymphocytes migrate across
the flat endothelium of a complex network of small vessels in the
marginal zone.(9) The MZ vessels receive an earlier supply of
arterial blood than the venous sinuses of the red pulp. What is
perhaps a more important distinguishing point than morphology
between the splenic and nodal selection mechanisms is a type of
experiment which the concept of which started with the work of
Gesner and Ginsburg.(25) When lymphocytes are treated in vitro with
enzymes or other agents intended to modify their cell surface in a
defined way, a perturbation in their migratory pattern may be
revealed by returning them intravenously to a syngeneic recipient.
This approach has been adopted by a number of groups in recent
years. Some of the results were summarized by Ford et al.(26) who
classified the agents as a) those which profoundly inhibited
migration into lymph nodes b) those which slightly but significantly
inhibited migration into lymph nodes c) those which induced the
prolonged retention of lymphocytes in the extravascular compartment
of certain tissues and d) those with no effect, (Table 1). It is

THE MIGRATION OF LYMPHOCYTES AFTER EXPOSURE IN VITRO TO VARIOUS REAGENTS

For references see (26) except for colchicine and cytochalasin B (27)

Reagent	Inhibition[1] of migration into — Lymph nodes	Spleen	Retention in specified organ	
Trypsin	+++	-	-	Entry into lymph nodes profoundly inhibited
Papain	+++	-	-	
Sodium azide	++	-	-	
Phospholipase A	+	-	Lungs	Entry into lymph nodes slightly inhibited
Phospholipase C	+	-	-	
Dextran sulphate	+	+|	-	
Cytochalasin B (27)	+	-	-	
Neuraminidase	-	-	Liver	Prolonged retention in certain tissues
Concanavalin A	-	-	All organs	
Colchicine (27)	+|	-	Lungs	
Bacterial lipopolysaccharide	-	-	Spleen	
Hyaluronidase	-	-	-	No effect
Thrombin	-	-	-	

1 +++ Inhibition to ≤ 10% of control
++ Inhibition to 10-50% of control
+ Marginal effect
- No effect

remarkable that, of seven substances known to inhibit the capacity
of lymphocytes to localize in lymph nodes, none substantially
affect migration into the spleen although dextran sulphate possibly
affects splenic localization marginally. Two substances
concanavalin A (28) and bacterial lipopolysaccharide (29) delay the
transit of lymphocytes through the spleen. Although both substances
are well known mitogens, the delaying effect of con A cannot be
attributed to blastogenic transformation because this is not
detected in vivo by administering ^{3}H-thymidine under the conditions
of these experiments.(28)

Another approach to the selective migration of lymphocytes is
to administer some agent to a recipient of labelled lymphocytes with
the intention of modifying the function of specialized endothelial
cells. Sulphated polysaccharides such as dextran sulphate induce a
lymphocytosis by interfering with the egress of lymphocytes from
the blood.(26,30) Although dextran sulphate has a slight effect
when it is applied to lymphocytes in vitro, there is a much greater
effect when recipients are treated, presumably because it acts
principally on vascular endothelium. In recipients treated with
dextran sulphate the early localization of syngeneic lymphocytes
in lymph nodes is reduced to about 11% of controls in the face of a
2-7 fold surplus in the blood. However, the proportion of injected
cells recovered from the spleen is the same as that in the spleens
of uninjected controls.(26) This result may have reflected the
balance of two opposing factors - the raised level of lymphocytes in
the blood tending to increase the numbers localizing in the spleen
and a partial inhibition of the capacity of lymphocytes to migrate
into the spleen due to dextran sulphate. An alternative
explanation - that a "lymph-node-seeking" subset is inhibited by
dextran sulphate but a "spleen-seeking" subset is unaffected was
disproved by secondary transfer experiments of labelled lymphocytes
in the blood of recipients which were untreated, dextran sulphate
treated or splenectomized. If separate "spleen-seeking" and "lymph-
node-seeking" subsets existed, then the former should be more
plentiful in the blood of splenectomized rats and the latter more
plentiful in the blood of dextran sulphate treated rats but the
results provided no support for this notion.(26) Therefore, the
view that the major population of recirculating lymphocytes may
enter the spleen or lymph nodes without organ preference remains
tenable.

There is now enough information on the agents which perturb
lymphocyte migration into lymph nodes to encourage speculation about
the mechanisms involved. For example, the inhibitory effects of
sodium azide and cytochalasin treatment on the adhesion of

lymphocytes to endothelial cells suggest that an active response by the lymphocyte involving its microfilaments is required.(26,31) By contrast, the inability to modify lymphocytes in such a way as to prevent selective migration into the spleen precludes any positive conclusions about the mechanisms involved. The lack of any effect of enzymes or lectins prompts the notion that physical rather than chemical factors may be involved. The possibility that rheological factors might contribute to the sorting out of lymphocytes in the blood vessels of the spleen deserves consideration.

LYMPHOCYTE RECIRCULATION AND THE IMMUNE RESPONSE

Lymphocyte recirculation achieves a continuous redistribution of lymphocytes to the scattered outposts of peripheral lymphatic tissues. This is thought to facilitate immune responses by permitting the gradual accumulation of rare antigen sensitive cells in a local depot of antigen and by increasing the chances of contact of infrequent antigen-specific T and B lymphocytes. In the spleen, T and B lymphocytes are channelled together as soon as they leave the blood in the marginal zone. They remain together for several hours in the PALS before segregating. If mutual contact or close proximity is required, then T-B cooperation probably occurs in the marginal zone or PALS. Both compartments include antigen-bearing macrophages, although their concentration is far higher in the marginal zone.(32) In the PALS, antigen-specific T cells are immobilized during the stage of blastic transformation. Conceivably each T cell sits in wait for the arrival in the nearby traffic stream of a B cell of corresponding specificity.

As for the significance of the later stages of B cell migration some vital clue or basic conception seems to be missing. Most B lymphocytes enter the corona around the germinal centre but we do not know what influences are brought to bear on them in that environment. Possibly there is a further stage of segregation into those cells destined to become memory cells and those which will become antibody forming cells. The route by which activated B cells reach the red pulp may be by short lymphatic channels or could be by the bloodstream by analogy with the route taken by IgA precursors activated in Peyer's patches and reaching the lamina propria of the gut after a short period in the blood.(33)

At present there is only a fragmentary understanding of lymphocyte recirculation and more information at the organizational level spanning the cell and the tissue compartment is necessary in order to gain further insight into the physiology of the spleen.

REFERENCES

1. RANNIE, G.H. and FORD, W.L. Physiology of lymphocyte recirculation in animal models. In: Transplantation and Clinical Immunology. Excerpta Medica, Amsterdam, Oxford, vol. IX. (1977).

2. GOWANS, J.L. and KNIGHT, E.J. The route of recirculation of lymphocytes in the rat. Proceedings of the Royal Society, Series B, 159: 257 (1964).

3. FORD, W.L. ET AL. The migration of lymphocytes across specialized vascular endothelium. II. The contrasting consequences of treating lymphocytes with trypsin or neuraminidase. Cell and Tissue Kinetics, 9: 351 (1976).

4. BRADFIELD, J.W.B. and BORN, G.V.R. The migration of rat thoracic duct lymphocytes through spleen in vivo. British Journal of Experimental Pathology, 54: 509 (1973).

5. FORD, W.L. The kinetics of lymphocyte recirculation within the rat spleen. Cell and Tissue Kinetics, 2: 171 (1969).

6. HAMMOND, B.J. A compartmental analysis of circulatory lymphocytes in the spleen. Cell and Tissue Kinetics, 8: 153 (1975).

7. FORD, W.L. The mechanism of lymphopenia produced by chronic irradiation of the rat spleen. British Journal of Experimental Pathology, 49: 502 (1968).

8. NIEUWENHUIS, P. and FORD, W.L. Comparative migration of B- and T-lymphocytes in the rat spleen and lymph-nodes. Cellular Immunology, 23: 254 (1976).

9. GOLDSCHNEIDER, I. and McGREGOR, D.D. Migration of lymphocytes and thymocytes in the rat. I. The route of migration from blood to spleen and lymph-nodes. Journal of Experimental Medicine, 127: 155 (1968).

10. GUTMAN, G.A. and WEISSMAN, I.L. Homing properties of thymus-independent follicular lymphocytes. Transplantation, 16: 621 (1973).

11. SEDGLEY, M. and FORD, W.L. The migration of lymphocytes across specialized vascular endothelium. I. The entry of lymphocytes into the isolated mesenteric lymph-node of the rat. Cell and Tissue Kinetics, 9: 231 (1976).

12. PABST, R. and TREPEL, F. The predominant role of the spleen in lymphocyte recirculation. I. Homing of lymphocytes to and release from the isolated perfused pig spleen. Cell and Tissue Kinetics, 8: 529 (1975).

13. FORD, W.L. The immunological and migratory properties of the lymphocytes recirculating through the rat spleen. British Journal of Experimental Pathology, 50: 257 (1969).

14. WEISS, L. The spleen. In: The Role of the Spleen in the Immunology of Parasitic Diseases (this volume).

15. MITCHELL, J. Lymphocyte circulation in the spleen. Marginal zone bridging channels and their possible role in cell traffic. Immunology, 24: 93 (1973).

16. McGREGOR, D.D. and GOWANS, J.L. The antibody response of rats depleted of lymphocytes by chronic drainage from the thoracic duct. Journal of Experimental Medicine, 117: 303 (1963).

17. STROBER, S. and DILLEY, J. Maturation of B lymphocytes in the rat. I. Migration pattern, tissue distribution and turnover rate of umprimed B lymphocytes involved in the adoptive anti-dinitrophenyl response. Journal of Experimental Medicine, 138: 1331 (1973).

18. STROBER, S. Maturation of B lymphocytes in the rat. II. Subpopulations of virgin B lymphocytes in the spleen and thoracic duct lymph. Journal of Immunology, 114: 877 (1975).

19. STROBER, S. Immune function, cell surface characteristics, and maturation of B cell subpopulations. Transplantation Reviews, 24: 84 (1975).

20. HALL, B.M. ET AL. The cellular basis of allograft rejection in vivo. 2. The nature of memory cells mediating second set heart graft rejection. Journal of Experimental Medicine, 148: (in press) (1978).

21. HALL, B.M. ET AL. Memory cells responsible for accelerated graft rejection do not recirculate in normal animals. Transplantation (in press) (1978).

22. CANTOR, H. and BOYSE, E. Regulation of the immune response by T-cell subclasses. Contemporary Topics in Immunobiology, 7: 47 (1977).

23. MACDONALD, H.R. ET AL. Generation of cytotoxic T lymphocytes in vitro. IV. Functional activation of memory cells in the absence of DNA synthesis. Journal of Experimental Medicine, 142: 622 (1975).

24. ANDERSON, A.O. and ANDERSON, N.D. Lymphocyte emigration from high endothelial venules in rat lymph nodes. Immunology, 31: 731 (1976).

25. GESNER, B.M. and GINSBURG, V. Effect of glycosidases on the fate of transfused lymphocytes. Proceedings of the National Academy of Sciences, 52: 750 (1964).

26. FORD, W.L. ET AL. Possible clues to the mechanism underlying the selective migration of lymphocytes from the blood. In: Cell-Cell Recognition. Cambridge University Press (in press) (1978).

27. SMITH, M.E. and FORD, W.L. The effect of cytoskeletal inhibitors on the migration of lymphocytes across vascular endothelium. In: Proceedings of the Sixth International Conference on Lymphatic Tissues and Germinal Centres in Immune Reactions. Plenum Press. (in press) (1978).

28. SMITH, M.E. ET AL. The migration of lymphocytes across specialized vascular endothelium. III. Concanavalin A delays lymphocytes in normal traffic areas. Experimental Cell Biology, 45: 9 (1977).

29. FREITAS, A.A. and DE SOUSA, M.A.B. Control mechanism of lymphocyte traffic. Altered distribution of ^{51}Cr-labelled mouse lymph-node cells pretreated in vitro with lipopolysaccharide. European Journal of Immunology, 6: 269 (1976).

30. BRADFIELD, J.W.B. and BORN, G.V.R. Lymphocytosis produced by heparin and other sulphated polysaccharides in rats and mice. Cellular Immunology, 14: 22 (1974).

31. WOODRUFF, J.J. ET AL. An in vitro model of lymphocyte homing. II. Membrane and cytoplasmic events involved in lymphocyte adherence to specialized high endothelial venules of lymph nodes. Journal of Immunology, 119: 1603 (1977).

32. NOSSAL, G.J.V. ET AL. Antigens in immunity. XII. Antigen trapping in the spleen. International Archives of Allergy, 29: 368 (1966).

33. HUSBAND, A.J. ET AL. The natural history of the cells producing IgA in the gut. Ciba Foundation Symposium 46 (New Series) p. 27 (1977).

ACKNOWLEDGEMENTS

Our unpublished work referred to in this paper was largely supported by MRC programme grant G972/455/B.

Specific Sequestration of Lymphocytes in the Spleen During the Immune Reponse Role of the Major Histocompatibility Complex

J. Sprent, Immunobiology Unit, Department of Pathology,
University of Pennsylvania, School of Medicine,
Philadelphia, Pennsylvania, USA

Lymphocytes in the spleen and lymph nodes (LN) consist of approximately equal proportions of T (thymus-derived) and B (marrow-derived) cells. The vast majority of these cells are long-lived lymphocytes which migrate continuously between blood and the central lymph via the lymphoid tissues.(1) The purpose of this article is to review briefly how exposure to specific antigen radically alters this pattern of blood to lymph recirculation.

NONSPECIFIC EFFECTS

Subcutaneous injection of antigen has a marked effect on the physiology of the draining LN. Within a few days the nodes increase rapidly in size and cellularity and there is a sharp increase in blood flow.(2) From the studies with ^{51}Cr-labelled lymphocytes, it has been shown that lymphocytes "home" much more readily to stimulated LN than to resting LN.(3) At one time this was considered to reflect a process of lymphocyte "trapping", i.e. that antigen stimulation nonspecifically impeded the rate at which lymphocytes traversed LN and gained access to the efferent lymphocytes. This notion is no longer tenable since direct measurements have demonstrated that there is no reduction in the speed of recirculation through stimulated LN.(4) The principal effect on cell traffic is a

Supported by grants AI-10961 and CA-15822 from the United States Public Health Services.

dramatic increase in the <u>numbers</u> of lymphocytes perfusing the node. This increase in perfusion rate is presumably a device for recruiting specifically-reactive lymphocytes from elsewhere in the body to enable these cells to respond locally to the antigen in question (see below). Whether a similar increased rate of cell traffic also occurs in the spleen, e.g. after intravenous injection of antigen, is not known.

SPECIFIC EFFECTS

One of the histological hallmarks of the immune response is an increased rate of lymphocyte proliferation in tissues harbouring the response. Since few lymphocytes are sessile, can one conclude that most of the proliferating lymphocytes have been mobilized from other tissues? If this were the case, the lymphoid tissues not participating actively in the response should be specifically depleted of lymphocytes reactive to the antigen concerned. A large body of evidence supports this prediction.

The simplest method for studying this question is to inject animals intravenously with an antigen, such as sheep erythrocytes (SRC), and then monitor the levels of specifically-reactive lymphocytes in the circulation, e.g. thoracic duct lymph, at various times thereafter. With this approach, studies in mice (<u>5</u>) and rats (<u>6</u>) have shown that the responsiveness of thoracic duct lymphocytes (TDL) to the injected antigen is abolished or greatly reduced within one day of antigen injection. This process of "negative selection" is specific (reactivity to other antigens is not affected) and there is no evidence that active suppression is involved. TDL regain their reactivity after 2-3 days and by five days the level of specific lymphocytes is markedly above normal ("positive selection"). This sequence of events affects both T and B lymphocytes and applies to a wide variety of antigens, including transplantation antigens.(<u>7</u>,<u>8</u>)

What happens to the reactive lymphocytes during this period? <u>A priori</u>, one would expect the cells to localize in regions where the antigen is concentrated. In this respect, when ^{51}Cr-labelled SRC are injected intravenously most of the radioactivity accumulates in the spleen (also the liver) and virtually none reaches the LN. Hence, the spleen is the predicted site for lymphocyte sequestration. Although there is now clear evidence that sequestration does indeed occur in the spleen, the initial experiments designed to prove this point gave unexpected results.(<u>9</u>,<u>10</u>)

The approach used was to study the adoptive immune response of lymphoid cells from mice injected intravenously with SRC one day previously. As shown in Table I, TDL from these mice gave virtually no response to the injected antigen but responded well to a different antigen, i.e. horse erythrocytes (HRC). Curiously, this state of specific unresponsiveness also applied to cells from the spleen, LN (Table I) and Peyer's patches (data not shown). These findings occurred when the recipients of the transferred cells were challenged with antigen on day 0 and plaque-forming cells (PFC) were measured on day 7. With antigen challenge on day 7 and measurement of PFC on day 14, by contrast, the response with spleen cells rose to normal or above normal levels whereas TDL and LN remained totally unresponsive.

These data imply that the spleen is indeed the main site of lymphocyte sequestration but that, for obscure reasons, the sequestered cells in the spleen are temporarily prevented from responding on adoptive transfer. Although the precise reason for this delayed response of spleen cells is not known, the explanation favoured is that the phenomenon reflects a transient inability of recently-stimulated lymphocytes to home to the required micro-enivironment on adoptive transfer. Support for this viewpoint has come from measuring spleen cell responses in vitro, i.e. where the problem of homing is avoided. Using a microculture system which measures precursor frequency, it was found in fact that 1-day-primed spleen cells had an increased level of precursor cells reactive to the injected antigen;(11) 1-day-primed TDL contained virtually no specific precursors. These findings are illustrated in Table II.

Direct evidence for splenic sequestration has also come from elegant studies of Ford et al. (12) on the localization of T cells involved in a systemic graft-versus-host reaction in rats. By measuring the extent of blood to lymph recirculation of homozygous T cells in F_1 hybrid rats compared to syngeneic rats, it was demonstrated that 10% of parental strain T cells were reactive to the transplantation antigens of the F_1 hosts. Radiolabelling techniques showed that these cells failed to enter the lymph and accumulated selectively in the spleen and, to a lesser extent, in LN. Here the cells proliferated extensively.

Although the spleen is the main site for lymphocyte sequestration after intravenous injection of antigen, the LN seem to take over this role when antigen is administered subcutaneously. Clear evidence for sequestration in LN came from studies in sheep in which a cannula was placed in the efferent lymphatic vessel of the popliteal LN.(13) After local subcutaneous injection of antigen,

TABLE I

ANTI-SRC RESPONSES BY TDL AND SPLEEN FROM MICE INJECTED WITH SRC ONE DAY BEFORE: RESPONSE MEASURED IN VIVO

Lymphoid cells tested	Antigen injected 1 day before	IgG PFC/spleen in irradiated mice at 7 or 14 days post-transfer			
		day 7		day 14	
		anti-SRC	anti-HRC	anti-SRC	anti-HRC
TDL	SRC	0	32,970(1.11)	0	28,920(1.21)
	HRC	102,680(1.07)	0	43,920(1.57)	0
Mesenteric LN	SRC	0	11,930(1.18)	0	22,490(1.10)
	HRC	19,100(1.22)	0	37,950(1.33)	20(4.0)
Spleen	SRC	110(3.63)	113,660(1.22)	70,450(1.05)	15,400(1.12)
	HRC	67,180(1.23)	20(3.0)	65,090(1.29)	36,460(1.03)

10^7 TDL or 3×10^7 spleen cells from mice given 10^9 SRC or HRC one day previously were transferred intravenously into irradiated (750R) syngeneic mice. The recipients were challenged with SRC and HRC (5×10^7 of each) either at day 0 (for day 7 PFC) or day 7 (for day 14 PFC) post-transfer; the latter group of mice received a small dose of marrow cells (10^6) on day 4 to prevent death from haemopoietic failure. The data show the geometric mean ($x/\div$ SE) of four mice per group. Only IgG responses are shown; IgM responses gave similar data.

Data from Sprent and Miller. (10)

TABLE II

ANTI-SRC RESPONSE BY TDL AND SPLEEN FROM MICE INJECTED WITH
SRC ONE DAY BEFORE: RESPONSE MEASURED <u>IN VITRO</u>

Lymphoid cells tested	Antigen injected 1 day before	Fraction of microcultures responding to SRC
TDL	SRC	0%
	HRC	50%
Spleen	SRC	99%
	HRC	72%

TDL and spleen cells from mice given 10^9 SRC or HRC
intravenously 1 day previously were plated in microcultures (7 x 10^4
lymphocytes/well) in the presence of SRC and irradiated (1200R)
spleen cells as "filler" cells. The fraction of wells producing
IgM anti-SRC antibody was measured 6 days later. Data from Sprent
and Lefkovits.([11])

the reactivity of the efferent lymph cells for the injected antigen
disappeared specifically after 1-2 days. Curiously, this also
applied eventually to cells collected from the unstimulated
contralateral node. The intriguing possibility that the latter
finding reflected a large scale mobilization of specific cells from
the unstimulated node to the stimulated node has yet to be proved.

In all of the above situations, the stage of negative selection
for the specific antigen lasts for only 1-2 days (slightly longer
in the popliteal LN model in sheep). Thereafter the progeny of the
cells proliferating in the lymphoid tissues re-enter the circulation
in large numbers as blast cells. When the precursor frequency is
very high, i.e. in responses to strong transplantation antigens,
positive selection is massive and specifically-reactive blast cells
can account for up to 90% of the cells entering the lymph after
4-5 days.

MECHANISM OF SELECTION

The sequence of events which leads to sequestration is still
poorly understood. The available data imply that in the case of

the spleen, which has a large blood supply, incoming lymphocytes are selectively withdrawn from the circulation as they encounter the antigen in question. Since most recirculating lymphocytes traverse the spleen at least once a day,(7) the entire recirculating lymphocyte pool (RLP) can be screened for specific lymphocytes over a short period. For immune responses occurring in LN, screening of the RLP is presumably less extensive since the blood supply to LN is comparatively small. As mentioned earlier, the sharp increase in blood flow occurring in LN during antigenic stimulation presumably potentiates the screening process.

A propos the mechanism of selection, the simplest notion is that lymphocytes bind the antigen concerned and become physically trapped, i.e. in the same way that, when lymphocytes are passed through an antigen-coated column in vitro, the specific lymphocytes remain bound to the column. The problem with this viewpoint is that binding to such columns only applies to B cells. Purified T cells pass unimpeded through antigen-coated columns and (with the notable exception of suppressor T cells) show virtually no capacity to bind free antigen in vitro.(14-16) How does one explain this paradoxical finding, particularly when it is borne in mind that T cells are no less specific for antigen than B cells? This is one of the most intriguing questions in current immunology. Broadly speaking, the answer seems to be that, unlike B cells, T cells recognize and respond to antigen only when it is presented in association with gene products of the major histocompatibility complex (MHC).

ROLE OF THE MAJOR HISTOCOMPATIBILITY COMPLEX IN SELECTION

T cells express a variety of different functions, e.g. they exert cell-mediated lysis against virus-infected target cells, collaborate with specific B lymphocytes in cooperative antibody responses, proliferate in response to antigen-pulsed macrophages in vitro, and control delayed-type hypersensitivity. Recently it has become clear that in all of these situations, MHC gene products play a decisive role in T cell stimulation and function.(17) The current view is that T cells do not recognize antigen per se but a complex of antigen bound to MHC determinants expressed on living cells. The key point here is that, with a few isolated exceptions, T cells and the cells presenting antigen have to share the same MHC determinants; antigen presentation by cells bearing foreign MHC determinants is nonstimulatory. This has led to the notion that the receptor repertoire of T cells is restricted to antigen presented is association with "self" MHC determinants.

Recent evidence suggests that MHC determinants also control selection to antigen in vivo. We approached this question by asking whether MHC determinants are important in the stimulation of T-helper cells, i.e. cells involved in T-B collaboration. Experiments in vitro have shown that for T-helper cells to express their function, the cells must first be activated to antigen presented by macrophages; the latter are required to be MHC-compatible with the T cells.(18) The prediction follows that if the same requirement applied in vivo, T cells exposed to antigen in vivo would presumably be stimulated, e.g. undergo selection, only if the macrophages presenting the antigen were MHC-compatible with the T cells.

To test this prediction we investigated whether selection to antigen occurred when macrophage-depleted T cells were exposed to SRC in an MHC-different environment.(19) The system was to transfer purified T cells plus SRC into heavily irradiated mice and then study the helper function of the donor cells recovered from thoracic duct lymph of the recipients one day later, i.e. during the stage of negative selection. (It should be mentioned that macrophages in irradiated mice are highly radioresistant.)

The data in Table III show that marked and specific negative selection occurred when syngeneic hosts were used for selection. Thus, when CBA strain T cells were transferred with SRC into irradiated CBA mice, the donor cells collected from TDL one day later stimulated CBA B cells to respond to HRC but gave no response against SRC. With T cells recovered at day 5, by contrast, above-normal responses to SRC were found.

Very different findings occurred when MHC-different hosts were used for selection. This is illustrated in the experiment shown in Table IV in which CBA T cells were selected to SRC in three congenic resistant strains; these mice all have identical "background" genes and differ only at the H-2 complex (the MHC of mice). Selection to SRC was profound when CBA (H-2^k) T cells were transferred with SRC to irradiated H-2 identical B10.Br mice. Significantly no selection occurred, however, when H-2-different B10 (H-2^b) or B10.D2 (H-2^d) mice were used for selection. Anti-SRC responses stimulated by the donor CBA T cells recovered from these mice were no lower than the responses given by T cells filtered in the absence of SRC. Similar findings also occurred when the T cells were harvested during the stage of positive selection (day 5), i.e. the helper function of the cells remained comparable to that given by T cells filtered without SRC (data not shown). (In this experiment it was necessary to use

TABLE III

NEGATIVE AND POSITIVE SELECTION OF CBA T CELLS TO SRC IN IRRADIATED SYNGENEIC MICE

T helper cells	SRC added during selection	Time of harvesting T cells from TDL	B cells	IgG PFC/spleen in irradiated CBA mice	
				Anti-SRC	Anti-HRC
CBA T cells	-	day 1-2	CBA	36,530(1.25)	44,670(1.16)
→ irrad. CBA		day 5-6	CBA	20,320(1.38)	18,700(1.16)
CBA T cells	+	day 1-2	CBA	0	36,620(1.45)
→ irrad. CBA		day 5-6	CBA	65,860(1.13)	24,480(1.07)

10^8 nylon-wool-purified LN T cells from mice primed with SRC + HRC 2 months previously were transferred intravenously ± SRC (0.5 ml of 50%) into irradiated (900R) CBA mice. Groups of these mice were cannulated either at day 1 or five days post-transfer and TDL collected overnight. 2.5 x 10⁶ of the lymph-borne T cells were then transferred with B cells (anti-θ-treated SRC + HRC-primed spleen cells) plus SRC and HRC into 700R CBA mice for measurement of PFC. Geometric mean of four mice per group. IgM PFC gave similar data.

Data from Sprent.(19)

TABLE IV

FAILURE TO INDUCE NEGATIVE SELECTION TO SRC IN IRRADIATED H-2-DIFFERENT MICE:
T CELLS HARVESTED FROM LYMPH AT DAY 1-2 POST-TRANSFER

CBA ($H-2^k$) T helper cells	$H-2$ region of filtration host	SRC added during selection	B cells	IgG PFC/spleen at 7 days in irradiated CBA mice	
				Anti-SRC	Anti-HRC
CBA T cells → irrad. B10.Br	$H-2^k$	–	CBA	45,300(1.11)	42,910(1.22)
CBA T cells → irrad. B10.Br	$H-2^k$	+	CBA	320(1.56)	47,130(1.19)
CBA T cells → irrad. B10	$H-2^b$	+	CBA	39,170(1.18)	37,070(1.15)
CBA T cells → irrad. B10.D2	$H-2^d$	+	CBA	48,100(1.11)	42,120(1.17)

The design of the experiment was identical to that shown in Table III. In the case of selection through B10 and B10.D2 mice, cytotoxic testing with appropriate alloantiserum plus complement showed that > 90% of the lymph-borne T cells were of donor CBA origin.

Data from Sprent.(19)

T cells which had first been depleted of specific alloreactive lymphocytes; without this precaution the cells entering the lymph at day 5 would have been mostly alloreactive blast cells.)

It is important to point out that selection <u>does</u> occur in the above system if F_1 hybrids are used for selection, i.e. there is no evidence that a concomitant graft-versus-host reaction impairs selection. It may also be mentioned that preliminary studies have shown that selection can be induced in <u>H-2</u>-different hosts if the donor cells are mixed with macrophage-enriched populations of donor origin. Hence, selection probably is indeed controlled by macrophages or related cells.

The main conclusion from these studies is that T cells fail to undergo either negative or positive selection to SRC on adoptive transfer unless the donor and host share <u>H-2</u> determinants. In an <u>H-2</u>-different environment, T cells ignore the injected antigen and continue their normal pattern of migration in an unstimulated state (Table V).

TABLE V

NEGATIVE AND POSITIVE SELECTION TO SRC DEPENDENT UPON SHARING OF <u>H-2</u> DETERMINANTS BETWEEN DONOR T CELLS AND IRRADIATED HOST

Strain combination	Day of collection from TDL	Adoptive response to:	
		SRC	HRC
T<u>a</u> + SRC → irrad. <u>a</u>	day 1	−	+
	day 5	+++	+
T<u>a</u> + SRC → irrad. (<u>a</u>x<u>b</u>)F$_1$	day 1	−	+
	day 5	+++	+
T<u>a</u> + SRC → irrad. <u>b</u>	day 1	+	+
	day 5	+	+

CONCLUDING REMARKS

Although other antigens and other T cell functions have yet to be studied, the above data strongly suggest that selection of T cell to antigen <u>in vivo</u> does not reflect contact with antigen <u>per se</u>.

Apparently the antigen must first be processed in some way, probably by macrophages, and then presented to T cells complexed to cell surface MHC determinants. Providing the antigen is complexed to self MHC determinants, the spleen screens the RLP for specific lymphocytes and selectively removes these cells from the circulation; once sequestered, the cells respond in situ to the complexed antigen. This chain of events fails to occur if the antigen is complexed to foreign MHC determinants. Splenic sequestration of T lymphocytes thus appears to obey the "rules" of MHC restriction laid down for a variety of other T cell functions. Whether sequestration of B cells is also under MHC control has yet to be studied.

REFERENCES

1. GOWANS, J.L. and McGREGOR, D.D. The immunological activities of lymphocytes. Progress in Allergy, 9: 1 (1965).

2. HAY, J.B. and HOBBS, B.B. The flow of blood to lymph nodes and its relation to lymphocyte traffic and the immune response. Journal of Experimental Medicine, 145: 31 (1977).

3. ZATZ, M. and LANCE, E.M. The distribution of ^{51}Cr-labelled lymphocytes in antigen-stimulated mice. Lymphocyte trapping. Journal of Experimental Medicine, 134: 224 (1971).

4. CAHILL, R.N.P. ET AL. The effects of antigen on the migration of recirculating lymphocytes through single lymph nodes. Journal of Experimental Medicine, 143: 870 (1976).

5. SPRENT, J. ET AL. Antigen-induced selective recruitment of circulating lymphocytes. Cellular Immunology, 2: 171 (1971).

6. ROWLEY, D.A. ET AL. The specific selection of recirculating lymphocytes by antigen in normal and preimmunized rats. Journal of Experimental Medicine, 136: 499 (1972).

7. FORD, W.L. Lymphocyte migration and immune responses. Progress in Allergy, 19: 1 (1975).

8. SPRENT, J. Recirculating lymphocytes. In: The lymphocyte: Structure and Function. (J.J. Marchalonis, ed.) Marcel Dekker, New York, p. 43 (1977).

9. SPRENT, J. and MILLER, J.F.A.P. Effect of recent antigen priming on adoptive immune responses. I. Specific unresponsiveness of cells from lymphoid organs of mice primed with heterologous erythrocytes. Journal of Experimental Medicine, 138: 143 (1973).

10. SPRENT, J. and MILLER, J.F.A.P. Effect of recent antigen priming on adoptive immune responses. II. Specific unresponsiveness of circulating lymphocytes from mice primed with heterologous erythrocytes. Journal of Experimental Medicine, 139: 1 (1974).

11. SPRENT, J. and LEFKOVITS, I. Effect of recent antigen priming on adoptive immune responses. IV. Antigen-induced selective recruitment of recirculating lymphocytes to the spleen demonstrable with a microculture system. Journal of Experimental Medicine, 143: 1289 (1976).

12. FORD, W.L. ET AL. Early cellular events in systemic graft-versus-host reaction. II. Autoradiographic estimates of the frequency of donor lymphocytes which respond to each Ag-B-determined antigenic complex. Journal of Experimental Medicine, 141: 681 (1975).

13. HAY, J.B. ET AL. The kinetics of antigen-reactive cells during lymphocyte recruitment. Cellular Immunology, 10: 145 (1974).

14. ELLIOT, B.E. ET AL. Thymus-derived rosettes are not helper cells. Journal of Experimental Medicine, 138: 1133 (1973).

15. BASTEN, A. ET AL. Relationship between Fc receptors, antigen-binding sites on T and B cells, and H-2 complex-associated determinants. Journal of Experimental Medicine, 141: 547 (1975).

16. TANIGUCHI, M. and MILLER, J.F.A.P. Enrichment of specific suppressor T cells and characterization of their surface markers. Journal of Experimental Medicine, 146: 1450 (1977).

17. KATZ, D.H. and BENACERRAF, B. The function and interrelationship of T cell receptors, Ir genes and other histocompatibility gene products. Transplantation Reviews, 22: 175 (1975).

18. ERB, P. and FELDMANN, M. The role of macrophages in the generation of T-helper cells. II. The genetic control of the macrophage-T cell interaction for helper cell induction with soluble antigens. Journal of Experimental Medicine, 142: 460 (1975).

19. SPRENT, J. Role of the H-2 complex in induction of T-helper cells in vivo. I. Antigen-specific selection of donor T cells to sheep erythrocytes in irradiated mice dependent upon sharing of H-2 determinants between donor and host. Journal of Experimental Medicine, in press (1978).

DISCUSSION

WEISS: The difference between Professor Ford on the one hand and
Dr Norman Anderson and me on the other is that we think there is a
significant flow of lymphocytes from the spleen through lymphatic
vessels and he thinks that there is not, that the lymphocytes exit
through the blood.

One line of evidence Professor Ford cites depends upon his
preparation of an isolated spleen perfused through its blood
vasculature. He shows that if he puts large numbers of lymphocytes
into the arterial end of such a preparation, even on a repeated
basis, they come out of the venous end and can be fully recovered.
I think that evidence from an isolated spleen, in which the
lymphatics have been cut, cannot speak to the role of lymphatics.
If there are two doors from a room and one is shut, everyone coming
in and going out must use the only open one. Similarly, that
lymphocyte traffic through lymph nodes is increased after
splenectomy is neither surprising nor does it speak to the question
of lymphatics of the spleen.

Judith Mitchell's work is excellent and it is likely that her
"bridging" zones may be a site for lymphocyte movement from white
pulp to red pulp. We did not find them to be so in our preparations.
Typically they contained a small arterial vessel with some
persistent periarterial lymphatic sheath which ran out into the
marginal zone.

Dr Anderson and I conclude that there are several pathways
by which lymphocytes may exit from the spleen. They may leave by
venous blood and, in fact, we find that soon after injecting
labelled lymphocytes most have left the spleen by the blood and can
continue to do so. A second route may be from the blood of the
splenic venous system into trabecular lymphatics and then out. But
we find that a significant number of thoracic duct lymphocytes in
our rats take about four hours to traverse the spleen and exit,
having moved across the periarterial lymphatic sheath, through the
deep lymphatics that entwine the central artery and its branch.
We have demonstrated a rich network of these efferent lymphatics, we
have shown they contain lymphocytes (see also Janont and Weiss,
Anatomical Records, 172: 197 (1972) and we predicate our conclusions
on the level of flow by cell counts in autoradiographs and
scintillation counts of portions of the spleen.

Our work has been submitted to Laboratory Investigation.

ALLISON: In adoptive transfer experiments spleen cells from immune
mice confer resistance to babesia infections much more efficiently

than do lymph-node cells. Is there any evidence about the properties of the subpopulation of lymphocytes that might be responsible for this immunity? Moreover, what is known about the T1 and T2 subpopulations of Canto and his colleagues in relation to circulation through the spleen?

FORD: One of the many criteria applied to distinguish T1 and T2 cells was the inability of T1 cells to recirculate from blood to lymph through lymph nodes. T1 cells were stated to be plentiful in the spleen but the histological site of these non-recirculating T cells was never defined. The application of anti-Ly sera to tissue sections may elucidate this question. There is good evidence for both B and T subsets in the spleen which do not belong to the recirculating pool. The possible roles of these subsets in immune responses are not clear.

MITCHELL: Professor Weiss, can you continue with this detailed account by telling us whether lymphatics emanating from the spleen pass through lymph nodes before entering the thoracic duct lymph? Secondly, is it correct that all splenic venous outflow passes to the portal system?

WEISS: In the rat there are 2-4 small lymph nodes situated in the mesentery between the thoracic duct and the spleen. Splenic veins empty into the portal system.

MITCHELL: With reference to the controversy concerning the role of the lymphatics in export of lymphocytes from spleen; when labelled lymphocytes are transfused to a cannulated rat, very few labelled cells appear in the thoracic duct. This suggests that there is no direct communication between splenic lymphatics and thoracic duct. However, if a lymph node is interposed and the transit time through the lymph node long, a delayed appearance in thoracic duct would be expected.

FORD: The argument I advanced hinges on the temporal pattern of recovery of intravenously injected lymphocytes from a thoracic duct cannula in a syngeneic recipient. Splenectomy before injection has been observed to have the opposite effect of what would be expected on the assumption that even a substantial minority, say 10%, of lymphocytes returning from the spleen do so via the thoracic duct.

Because of quantitative considerations, the lymph nodes in the vascular pedicle of the spleen are unlikely to complicate the situation. In a rat of 200g the spleen is accepting and releasing about 80×10^6 lymphocytes per hour. The tiny lymph nodes draining the spleen weigh only a few mg. - say a maximum of 4mg. If there were as many as 10^6 lymphocytes per mg. of lymph node en route from afferent to efferent lymph in these nodes then this would delay 8×10^6 lymphocytes by only 0.5 hours. This would not affect the argument which is based on an average lymph node transit time of 15–18 hours for T lymphocytes in contrast to a splenic transit time of 4–5 hours.

FORD: Dr Sprent, some of your experiments provide new and persuasive evidence of the requirement for macrophage processing in the antibody response to sheep erythrocytes. Would you please elaborate on the experiment in which you restored the partition of antigen-reactive lymphocytes by providing macrophages syngeneic with the lymphocytes?

SPRENT: These experiments are still very preliminary but they suggest that if the donor T cells are mixed with H-2 compatible macrophage-enriched populations, i.e. T-cell-depleted peritoneal exudate cells, selection can be induced in H-2-different irradiated mice. This effect is not seen with macrophage-depleted populations, e.g. B cells from thoracic duct lymph. Hence the antigen-presenting cell is probably a macrophage-like cell. Its precise identity, however, is far from clear.

COHEN: At the time when sensitized lymphocytes re-enter the circulation, is the spleen completely depleted of such cells?

SPRENT: In the case of normal mice injected with SRC the spleen shows a moderate increase in reactivity at day 1-3 post-transfer and a marked increase by day 5. At no time point examined is there a decreased response.

BATTISTO: Have you considered the possibility of specific suppression that might be expressed by SRBC in contact with the T cells at day 1? Has any attempt been made to eliminate suppressor T cells from the transferred cells?

SPRENT: I think that suppression is an unlikely explanation for two reasons. Firstly, addition of the selected T cells to normal T cells does not inhibit the function of the latter. Secondly, the unresponsiveness lasts only a very brief period (1-2 days) and is rapidly followed by a period of marked hyper-reactivity.

CLAYTON: I believe that your nylon columns would have removed your suppressor cells.

SPRENT: Yes, there are indeed claims that nylon wool columns remove suppressor cells. This might be taken to reinforce the argument that the negative selection observed at day 1 post-transfer is extremely difficult to explain in terms of suppression.

MILLER: Dr Sprent, your last conclusion has profound implications for the mechanism by which the T cell repertoire is generated and selected. It appears from the very clear data you presented that the reactivities of T cells are directed to antigen only in association with self-MHC components and is imposed before priming. Is it possible, however, that your T cells were primed to environmental antigens which cross-react with sheep or horse erythrocytes so that the restriction you observed is imposed as a result of priming and is not pre-existent? If not, how do you account for the results of several investigators, e.g. Thomas and Shevach, who showed that T cells can be primed to antigen presented on allogeneic macrophages?

SPRENT: You raise a very important question. I agree that one cannot exclude the possibility that mice are naturally primed to SRC as the result of environmental priming. But the same criticism might be levelled against virtually any antigen, even a synthetic antigen. Nevertheless one clearly can't generalize from the data I've presented until other antigens have been studied. However I think the strongest evidence for T cells being H-2-restricted before priming comes from the finding that F_1 hybrid T cells raised from stem cells in irradiated parental strain mice are unable to interact with cells bearing H-2 determinants of the opposite parental strain. (see Immunology Review, 42 (1978). As you know, these and other data strongly imply that H-2 restriction is imposed early in T cell differentiation as the result of confrontation with H-2 determinants on thymic epithelial cells. However, although most of the available data are consistent with this view point, you rightly point out that certain data, e.g. those of Thomas and Shevach, is against this notion. Just what these "anomalous" data signify is impossible to say at the present time.

FRANK: Have you studied soluble antigens?

SPRENT: No.

Splenic Regulation of Humoral and Cellular Immunological Responses in Other Domains

W.L. Welles and J.R. Battisto,
Department of Immunology, Research Division,
Cleveland Clinic, Cleveland, Ohio, USA

INTRODUCTION

Immunological competence is the result of a carefully regulated interplay of the cellular and molecular mechanisms of surveillance, tolerance and response. In recent years, as knowledge of these mechanisms has evolved, the spleen has emerged as an important regulatory organ with multi-dimensional impact. In addition to being the main site of immunoglobulin and antibody synthesis, data have shown that the spleen affects that immune capability of cellular populations in other lymphoid organs. This concept of the spleen as a lymphoid organ that is influential upon other immunological domains, first proposed several years ago (1) and recently reiterated,(2) will be re-examined by the information to be presented within this chapter.

Some understanding of splenic impact upon immunological function has been derived from reports of human splenectomy (3) and congenital asplenia.(4,5) However, until recently, human splenectomy has been indicated primarily following trauma or for management of warm-antibody autoimmune haemolytic disease.(6) Only now is interest expanding to include the consequences of splenectomy for patients with malignant disease.(7)

A systematic examination of splenic regulation of various immunological parameters has been pursued in laboratory studies with both hereditarily asplenic (8,9) and splenectomized mice. The combined use of animal models which are hereditarily asplenic and those which are surgically splenectomized has provided information about the temporal aspects of splenic regulation. Since understanding of the effects of asplenia upon immunological integrity has been most advanced by studies in the laboratory mouse, and since in-depth examination of the splenic role in disease is presented by numerous other authors within this text, this chapter will focus primarily on knowledge of the multifaceted roles of splenic lymphocytes in immunity as derived from murine studies.

PLASMA PROTEIN CONCENTRATIONS

Historically, the contribution of the spleen to plasma protein levels has been determined following its surgical extirpation. However, such characterization of the splenic role might, at best, be skewed or incomplete since it precludes any estimation of splenic influences at various levels during presurgical stages of growth. A more complete, and perhaps more valid, assessment is possible in studies of animals with total splenic agenesis such as the Dh/+ mouse.

Protein analysis of plasma from hereditarily asplenic mice (10, 11) has indicated that, although absolute protein levels are depressed (87% of normal), the relative percentages of albumin, alphaglobulins, betaglobulins and gammaglobulins remain normal. Measurements of individual serum immunoglobulin classes in asplenic (Dh/+) and eusplenic (+/+) littermate mice have shown considerably different values in two laboratories.(12,13) Fletcher et al. (12) have reported that hereditarily asplenic mice possess IgG_1, IgG_2, IgM and IgA at concentrations similar to those found in normal mouse strains.(14) In contrast, Fiocchi (13) has reported changes in the relative proportions of the different immunoglobulin classes and significantly higher levels of each immunoglobulin class as well These latter data are in general agreement with concentrations of total gammaglobulin levels reported earlier (11) for Dh/+ mice. An explanation for the differences in the absolute amounts as well as the relative percentages of the separate immunoglobulin classes in hereditarily asplenic mice is not yet available. The general observation which can be made from both of these studies is that the amount of immunoglobulin in the asplenic mouse is decreased relative to the normal littermate.

In the human situation, splenectomy of patients with Hodgkin's disease and other malignant lymphomas (7,15) has shown no alteration in the levels of serum IgG and IgA but significant decreases have occurred in IgM. Analogous results have been noted for both neonatally- and adult-splenectomized mice.(16,17) Serum levels of IgG_1, IgG_{2a}, IgG_{2b}, and IgA are largely unaffected by splenectomy but levels of IgM are reduced by 40-50%.[17] There appear to be, therefore, consistent changes produced by removal of the spleen in both the human and murine species.

ANTIBODY PRODUCTION

Although its function in humans is poorly understood, the spleen is centrally involved in the clearance of particulate antigens through production of opsonins which facilitate phagocytosis and elaboration of specific non-phagocytic antibody responses. Indeed, the spleen is the primary filter in the vascular system, sequestering bacteria introduced into the bloodstream. Furthermore, the phagocytic function of polymorphonuclear leucocytes is promoted by a bioactive tetrapeptide, termed tuftsin, which is released in the spleen following enzymatic digestion of a specific cell-bound IgG.[18] The deficiency of this particular factor has been characterized by onset of an extremely abrupt and fulminant bacterial infection such as noted in post-splenectomy septic episodes.[19] This grave hazard of septicaemia following splenectomy [20] has provided the impetus for re-evaluation of the criteria which mandate the surgical procedure.

Since the mouse spleen is known to be the major site of immunoglobulin synthesis up to approximately six months of age,[21] numerous investigators have studied the scope of splenic effects upon antibody synthesis by examining the ability of hereditarily asplenic mice to produce antibody in response to specific antigenic stimulation. In the first experiments, asplenic mice were challenged by immunization with bacteria and results indicated that, despite absence of the spleen, antibody was produced in detectable amounts.[10] Further quantitation of synthesis in Dh/+ mice [22] showed that serum levels of both the 7S antibodies, which mediate interspecies passive cutaneous anaphylaxis (IgG_2), and the 19S haemagglutinins (IgM) were depressed while levels of IgG_1, the antibody which mediates passive cutaneous anaphylaxis within a species, appeared normal. The depressed levels of IgM synthesis found in asplenic mice in vivo correlate with the fact that the majority of plasma cells which synthesize IgM are located in spleens of normal animals.[21]

Studies of lymph node antibody responses in vitro have further confirmed the in vivo data that lymph-node cells do not functionally replace the spleen in its absence.[23,24] Data in Table I show that lymph-node cells from Dh/+ mice respond at a level equal to or less than that of normal mice. In fact, the data also indicate that

TABLE I

LYMPH-NODE CELLS FROM NEONATALLY SPLENECTOMIZED
AND SOME HEREDITARILY ASPLENIC MICE RESPOND
DEFICIENTLY TO PRIMARY IMMUNIZATION IN VITRO WITH
SHEEP ERYTHROCYTES

Source of LN cells	No. of cultures	% Mean viability ($\pm$ S.E.)	Direct PFC*/10^6 ($\pm$ S.E.)	Mean DPFC/10^6 ($\pm$ S.E.)
CBA/J	3	22 (7)	118 (6)	94 (13)
	2	37 (2)	92 (12)	
	2	24 (3)	73 (0)	
CBA/J, neonatally splenectomized	3	24 (4)	19 (6)	24 (7)
	4	41 (1)	40 (6)	
	3	46 (5)	9 (9)	
	4	28 (4)	30 (4)	
+/+ (NORMAL)	2	22 (0)	118 (39)	104 (14)
	4	38 (2)	90 (13)	
Dh/+ (ASPLENIC)	6	30 (2)	112 (5)	86 (26)
	4	24 (1)	60 (14)	
	3	28 (6)	20 (11)	29 (9)
	4	20 (3)	38 (4)	

* Plaque-forming cell

lymph-node cells of noenatally splenectomized mice have a much
lower capacity for IgM synthesis than lymph-node cells of both
normal and hereditarily asplenic mice. An interpretation
consistent with these data is that the spleen exerts a regulatory
effect upon the antibody synthesizing capability in the lymph nodes
and perhaps other lymphoid sites as well. That is, lymph-node
cells which have never been exposed to splenic tissue (as in Dh/+
mice) are able to function almost normally, whereas lymph-node
cells which have been exposed to splenic tissue during periods of
fetal and neonatal development are not capable of providing a
normal response. Whether this is attributable to a permanent
suppressive effect produced by the spleen prior to its surgical
removal is a question that has yet to be investigated. Possibly a
molecular splenic signal causes the production of lymph node
suppressor cells in situ or perhaps splenic suppressor cells or
suppressive factors localize in lymph nodes at some pre-surgical
stage and remain following splenic extirpation.

Extensive studies of antibody synthesis in normal, splenectomized and asplenic mice have been reported by Lozzio and Wargon.(25,26) They have examined development of both IgM and IgG in the primary response of peripheral blood and lymph-node cells and show that the kinetics of antibody production are strongly regulated by the spleen. They have also studied primary and secondary responses to lipopolysaccharide and sheep erythrocytes and report elevations in IgG at the expense of IgM. Study of secondary antibody responses by measurement of IgM and IgG plaque-forming cells in bone marrow, lymph nodes, peripheral blood and spleen (where possible) of normal, asplenic and splenectomized mice has shown that asplenic mice produce less antibody than normal animals but are more competent in this capacity than splenectomized mice. In both primary and secondary responses, the asplenic mouse possesses antibody function at levels nearer to normal than the splenectomized mouse. These results further support the concept that the spleen exerts suppressor influences upon extra-splenic lymphoid organs.

Since one of the major functions of the spleen is antigen-trapping, marked alterations in this capacity might be expected in the asplenic condition. Profound changes attributable to asplenia were first revealed through the randomized antigen distribution found in Dh/+ mice.(27,28) Further studies (29) have shown that Dh/+ mice retain less than half of the radiolabelled sheep erythrocytes found in normal mice. This loss is due not only to absence of that fraction of antigen which would normally localize in the spleen but is augmented by depressed antigenic retention in the lymph nodes of asplenic mice. The absence of the spleen, therefore, significantly alters the antigenic stimulation experienced by at least one of the remaining lymphoid organs.

There are a number of important facts to be gleaned from the considerably complex results of the various studies of antibody synthesis in asplenic individuals. First, antigenic trapping is seriously impaired. Second, the kinetics of antibody production are considerably altered. Third, changes in antibody levels following neonatal splenectomy are distinct from those evident in situations of hereditary splenic agenesis. Last, and possibly most significant, the remaining lymphopoietic tissues do not compensate for absence of the spleen during embryogenesis nor does a neonatal spleen transfer restore immunocompetence for antibody synthesis to the hereditarily asplenic mouse.(26)

LYMPHOCYTE DYSFUNCTION

Much recent work has centred on studying the rate of appearance and number of murine lymphocyte populations in lymphoid tissues from fetal to early adult life.(30,31) From the 15th or 16th days of gestation, when B and T cells first appear, both populations increase in the mouse spleen to reach a plateau by one to two weeks of age.(32) Although T and B cell numbers are fully developed in the young mouse spleen, humoral antibody responses remain depressed for several weeks after birth.(33) Primary immune responses to several thymic-independent antigens, pneumoccocal polysaccharide Type I, Type II and lipopolysaccharide (LPS), are well developed by seven to 14 days of life. In contrast, adult primary and secondary responses to the T-dependent antigen, sheep erythrocytes, are only achieved by four to eight weeks of age.(34) Thus, T-cell responsiveness appears to mature to full expression during postnatal T-cell residency in the spleen.

What effects, if any, does hereditary or surgical deletion of the spleen have upon the quantitative features of cell populations in other lymphoid organs? Further, are T and B cell populations qualitatively different in the absence of the spleen? Numerous experiments have been designed to detect and evaluate such changes through use of the asplenic mouse.

The quantitative effects of hereditary asplenia upon cellular populations in the lymph nodes and thymus are not clear. There exists reports of normal (11) and enlarged thymi (12) as well as data for normal (12) and hypertrophic lymph nodes.(28) Thus far, experiments have shown that equal numbers of thymic T cells in both normal and asplenic mice bear the T-cell antigen, theta, and that equal number of macrophages exist in the lymph nodes of these animals.(12)

Splenic influence on functional lymphocyte maturity in antibody synthesis has been under examination since 1971 when Battisto et al. (1) reported that bone-marrow (BM) and thymic (T) cells of Dh/+ mice (asplenic) were incapable of expressing synergy for antibody production. As shown in Table II, reconstitution of normal, irradiated mice with bone marrow and thymus cells from +/+ mice (eusplenic) resulted in a competent primary IgM antibody response against the T-dependent antigen, sheep erythrocytes, but reconstitution with bone marrow and thymus cells from Dh/+ mice did not. Use of +/+ T cells or +/+ BM cells with Dh/+ BM cells or Dh/+ T cells, respectively, indicated that neither cell population of Dh/+ mice was capable of cooperating with the counterpart from normal mice. To obviate the possibility that cells of Dh/+ mice

TABLE II

BONE-MARROW AND THYMUS CELLS FROM HEREDITARILY
ASPLENIC MICE (Dh/+) DO NOT COOPERATE FOR
ANTIBODY SYNTHESIS

Type of syngeneic cells transferred (10^7 each)		No. of irradiated +/+ recipients	Mean Direct PFC/10^6 $\pm$ S.D.
BM	Thymus		
+/+	none	5	8.6 ± 1.5
Dh/+	none	4	7.0 ± 4.6
+/+	+/+	7	76.0 ± 4.6
Dh/+	Dh/+	15	8.8 ± 4.9
Dh/+	+/+	13	8.4 ± 4.6
+/+	Dh/+	11	7.6 ± 3.8

were unresponsive due to genetic defects or histocompatibility
barriers, the experiments were repeated with splenectomized mice of
the fully inbred CBA/J strain. Splenectomy of CBA/J mice at birth
was shown to produce a similar loss of cellular cooperation;
however, for splenectomized mice, the deficit appeared restricted
to the T-cell population.(35) Additional studies with asplenic
mice showed that their deficiency was also localized in the T cell.
Although the T cell was labelled defective, this term is not meant
to be definitive and includes the possibility that the T cell is
merely immature and not permanently dysfunctional. Additionally,
but less likely, the deficit could represent a total absence of the
assisting T-cell population. Several procedures were found to be
effective in restoring cellular synergy to splenectomized animals.
Delay of spleen removal until day 6 after birth caused a 60%
restoration of the response. Intraperitoneal infusion of autologous
spleen cells immediately after splenectomy or isogeneic spleen cell
infusion two months post splenectomy were both found to be totally
restorative. These data indicate that, although the spleen has
pronounced effects on B and T cells in the first days of life, it
can also strongly affect cellular function when introduced into the
adult mouse.

The ability of lymphocytes from neonatally-splenectomized mice to acquire the capacity for cooperation can be promoted within an irradiated splenic environment.(36,37) IgM responses of bone-marrow and thymus cells transferred from adult neonatally-splenectomized mice to irradiated adult hosts develop to normal levels after 14 days. The same recovery has been found for IgG antibody synthesis after three weeks and three immunizations with antigen.(36) These data support the view that T-helper cells of asplenic and splenectomized mice are maintained in a naive state and can be prompted by the splenic environment and repeated antigenic exposure to acquire the cooperative ability needed for antibody synthesis. That T and B cells required for antibody synthesis are maintained in a naive form due to arrested maturation in mice without spleens has been confirmed by additional data.(24, 37,38) However, in contrast to neonatally splenectomized mice, transfer of spleen cells to hereditarily asplenic mice does not restore synergy to their B and T cells.(1,37) This may indicate that splenic agenesis produces a germinal cell defect whereas neonatal splenectomy is less severe in that cells, particularly T, which have been exposed to splenic influences prenatally, can be stimulated to further differentiation upon transfer into a new environment.

Fletcher, et al. (12) have reported an interesting observation on hereditarily asplenic mice. According to these authors the diminished synergy for antibody synthesis in young, asplenic mice is attributable to T cells and is corrected with age. They also noted that as young adults (six weeks) these animals display a defective T-helper cell capacity for antibody synthesis, but they further discovered that by five months of age bone-marrow and thymus cells from hereditarily asplenic mice are as capable of primary responses to sheep red blood cells as cells from normal littermate mice. These data indicate that there may be an alternate, albeit slower, maturational process which can function in the asplenic mouse to promote differentiation of T-helper cells for antibody synthesis.

Although the defective synergy of B and T cells from asplenic mice has been confirmed in several laboratories (12,37,38) there are other reports which have questioned the presence of a defective T-helper cell for antibody synthesis.(17,24,39) Rozing et al. (17) deny the existence of a defect in the T-helper cell of neonatally splenectomized mice and attribute the observation of depressed antibody production, as seen by other investigators, to a suboptimal test system with aberrant sensitivity. However, the data of Fletcher et al. (12) clearly refute such an explanation since the same assay detects both defective and competent T-cell function in animals of different ages. The amassed data from different laboratories now confirm that the lack of synergy for antibody synthesis is not an error of protocol but a true result of asplenia.

Further effects of the spleen upon T cells have become evident through mitogenic and homing experiments. As shown in Table III,

TABLE III

SPLENIC REGULATION OF BONE-MARROW AND
LYMPH-NODE CELLS/ RESPONSES TO PHA-M

Type of B6.CBA mouse	PHA-M	Bone-marrow cells		Lymph-node cells	
		CPM at 72 hr.	Stimulation index	CPM at 72 hr.	Stimulation index
+/+	none	763	2.1	211	6.5
	10μl	1625		1371	
Dh/+	none	161	0.8	201	21.0
	10μl	133		4241	

bone-marrow cells of Dh/+ mice do not respond to stimulation by phytohaemagglutinin-M (PHA-M) when compared to normal bone-marrow cells. However, lymph-node cells of asplenic mice exhibit a stimulatory response to PHA-M which is three-fold larger than that of lymph-node cells from normal animals.(40) Since the precise significance of PHA responsiveness to cellular maturity and function has not yet been clarified, it is not possible to describe the transformation involved, but the data do indicate that lymphoid cells of bone marrow and lymph nodes are significantly different in the absence of the spleen. Thus, the spleen appears to mediate significant changes upon certain subpopulations of T cells in other lymphoid organs.

Additional data of splenic effects on T cells have been reported by Russell and Golub.(38) They have shown that thymic T cells from hereditarily-asplenic and neonatally-splenectomized mice respond to stimulation by the B-cell mitogen, lipopolysaccharide (LPS). Although the LPS response has not been confirmed,(12) further evidence for splenic effects on thymi are seen in the homing of radiolabelled thymus cells. Thymocytes from asplenic (12,38) and neonatally-splenectomized (38) mice exhibit a significant decrease, compared to normal thymus cells, in their ability to localize in spleens of normal animals. In contrast to the effect of spleen cell transplants on bone-marrow and thymus synergy, neither normal nor irradiated spleen grafts have any apparent capability to correct the defect in homing. Since BM and T-cell synergy for antibody synthesis can be recovered without restoration of normal T-cell homing patterns, the deficient homing of thymus cells from asplenic mice to the spleen can be excluded as a major factor contributing to the altered BM and T cell synergism seen in spleenless mice. Furthermore, homing may be an age-dependent phenomenon since the thymocytes of Dh/+ and +/+ mice that are six months old home to the spleen in identical percentages.(12) As stated earlier, by five months of age hereditarily-asplenic mice display normal cooperation between their BM and T cells for anti-sheep erythrocyte antibody synthesis. The concurrent change implies that the same differentiation or maturation process may be affecting both of these apparently unrelated T-cell populations.

To date, the control of the spleen on thymocyte, bone-marrow and lymph-node cell populations is not a fully defined area. Without doubt, the spleen has a major effect upon helper-T cells for antibody synthesis. The defect apparently diminishes with age in the Dh/+ mouse and the response is normal by 5-6 months of age. This correlates well with the information that the spleen is the main site of antibody synthesis up to six months of age. In older mice, the relative contribution of bone marrow increases, possibly due to a gradual shift in the individual animal from primary type responses to secondary type responses.(21)

CELLULAR IMMUNITY

Cell-mediated immunological integrity is provided by various populations of T cells. Responses included in this category are delayed type hypersensitivity, allograft rejection and the graft-versus-host (GVH) reaction. The in vitro measurement of T-lymphocyte response to allogeneic stimulation is accomplished by two procedures known as the mixed lymphocyte reaction (MLR) and cell-mediated lympholysis (CML).

Several lines of experimentation have indicated that at least some populations of T cells in hereditarily-asplenic mice function in a normal manner. The responses which have been found to be normal include contact sensitivity to a chemical allergen.(23) induction of a tolerant state to the same chemical hapten,(23) and expression of conventional primary and secondary allogeneic skin graft rejections.(1) Various laboratories have confirmed and extended these data.(12,25,41) In addition, an alcoholic extract of calf spleen, with the in vivo characteristics of a lymphoid chalone, has been shown to prevent or control an acute GVH reaction in lethally-irradiated mice. (42) The immunosuppression obtained is apparently specific for immunologically-competent grafted cells since the grafted haemopoietic stem cells remain unaffected. This observation indicates that a splenic substance can protect the host by controlling immunologically aggressive cells of the graft. Since this system might have importance for bone-marrow transplantation, further studies are indicated.

The influence of the spleen on proliferative responses of T cells from neonatally-splenectomized mice to stimulation by allogeneic cells has also been examined. (24) Data indicate that both lymph-node and thymus cells of splenectomized mice are as competent to respond to allogeneic stimulation (^{3}H-leucine uptake) by histocompatibility antigens as are cells from control littermates. Identical studies with lymph-node and thymus cells of Dh/+ mice also show no alterations in T-cell responder capability as measured in vitro by MLR. These data, which show unaltered proliferation to allogeneic stimuli by cells of asplenic mice, contrast with the results in the previous section which indicate si nificantly increased proliferative responses of lymph-node T cells to the mitogen PHA-M. Comparison of the results may lead to either of two interpretations: absence of the spleen selectively alters certain responses of a T cell (e.g. PHA-M responses but not allogeneic responses) or that separate and distinct subpopulations of T cells provide each of these responder capabilities and only one but not the other T cell subpopulation is affected by a spleenless environment.

Streilein and Wiesner (43) have recently reported a striking distinction in the responses of normal and adult-splenectomized female mice to weak histoincompatibility provided by the male H-Y antigen. Females with intact spleens responded to allografts of male skin by generating within their lymphoid tissues increased numbers of cells capable of mediating a GVH reaction when transferred to male mice. In contrast, the lymphoid systems of females without spleens appeared to be enriched for cells capable of mediating acute skin-graft rejection but were devoid of cells which, following transfer to male mice, proliferate in a manner indicative of GVH. That is, lymph-node cells of splenectomized, specifically-sensitized female mice, when transferred into normal male animals, do not manifest the GVH reactivity which is seen for specifically-sensitized female mice possessing spleens. The results indicate that, in this system of weak antigenic disparity, the spleen has two effects. First the spleen provides a strong suppressive influence which delays graft rejection and, second, the spleen is essential for development of a lymphoid-cell population

which proliferates in response to stimulation by host H-Y antigen.
The first point is reinforced with data recently reported on
rejection of male skin grafts by splenectomized female mice.(44)
Female mice of the C3H strain do not, under normal circumstances,
reject skin grafts from syngeneic males. However, 40% of
splenectomized C3H female mice completely rejected C3H male skin
grafts applied two weeks post splenectomy and all grafts on
splenectomized females showed at least transient signs of crisis.
The mechanisms involved in splenic regulation of effective or
abortive homograft rejection are currently unclear.

Suppressor cells and suppressor activity have been areas of
intense immunological interest in recent years.(45,46) At present
there exists a great deal of phenomenological evidence that
suppressor activity, whether cellular or molecular, influences the
magnitude and quality of diverse immune responses.(47,48,49) Thus,
the stimuli, generation and mechanisms of suppression are undergoing
critical study and evidence is accumulating to indicate that the
spleen is an essential organ for suppressive activity.

Cells of the neonatal spleen have been found to have a
suppressive effect in antibody synthesis by B cells of any age.(50)
The predominant functional activity of neonatal splenic T
lymphocytes appears to be a suppressive influence which maintains
the newborn animal in a state that is unresponsive to antigenic
stimulation. Additional data on the presence of suppressor
capability in the spleen of the neonatal mouse has been derived
from mixed leucocyte cultures with allogeneic cells.(51,52) Results
indicate that newborn spleen cells or soluble spleen-cell products
are able to suppress stimulation of responder adult lymphocytes.
Thus, there appears to be a suppressive regulatory function in the
neonatal spleen which affects replication in mixed leucocyte
cultures in a manner comparable to the T suppressor cell for
antibody synthesis. The high suppressor cell activity in the spleen
of neonates may have important implications for establishment of
self-tolerance. Indeed, high levels of suppressor activity are
known to exist during some experimentally induced states of
immunological tolerance.(53,54)

Knowledge in the area of splenic suppression has advanced from
studies of tolerance to contact sensitization with haptens. The
first experiments to test delayed hypersensitivity, a T-cell
response, in the asplenic mouse established that the spleen was not
essential to the induction of sensitivity or of tolerance to picryl
chloride.(23) Recent work with splenectomized mice (41) has
extended the understanding of mechanisms which establish the
tolerant state. Data show that splenectomized mice can be made

fully tolerant to challenge with dinitrofluorobenzene. However, unlike lymph nodes of tolerant mice, those of tolerant splenectomized animals are phenotypically tolerant and are unable to transfer tolerance to naive recipients (see Table IV),

TABLE IV

TRANSFER OF TOLERANCE BY LYMPH-NODE CELLS
FROM TOLERANT MICE *

Treatment	Transferred cells x 10^6	% Tolerance transferred
Sham-Splx	100	61
Sham-Splx	140	59
Sham-Splx	200	58
Splx	100	12
Splx	140	11
Splx	200	8

* Data of SY, M. ET AL. Journal of Immunology, 119: 2097 (1977).

Therefore, generation of a lymph-node suppressor T cell, which mediates transferrable tolerance but is not essential to phenotypic tolerance, can only occur in the presence of the spleen. Kinetic studies have shown that the spleen must be present at least three days after tolerization in order to permit development of a significant number of suppressor cells in the peripheral lymph nodes. Tolerogens in this system appear to stimulate either a cellular or a molecular suppressor capability in the spleen which then travels to the peripheral lymph nodes. These studies provide the first evidence that, of the three main categories of splenic suppressor cells: T cells,(46,47) B cells (55) and macrophages,(56) at least one suppressor population cannot develop in the absence of the spleen.

Examination of the complete spectrum of splenic roles in suppression is beyond the scope of this paper. However, two additional studies provide noteworthy information. Gershon et al. (57) have presented evidence that spleen-localizing thymocytes exert a suppressive effect on the DNA-synthesizing response to antigen of cells which localize in the lymph nodes. Removal of the spleen-localizing thymocytes by splenectomy three hours post inoculation markedly increased the amount of DNA synthesized in the lymph nodes. Furthermore, removal of the spleen-seeking thymocytes and reinoculation into a third mouse resulted in the re-appearance of the suppressive ability. The second study (2) indicates that there exist suppressor cells for antibody production and that, once activated, these cells may gain the ability to migrate from the spleen into the circulation and be disseminated to other lymphoid tissue. Both of these studies impart a commanding regulatory function to the spleen.

DIFFERENTIATION OF CELL SURFACE ANTIGENS

The importance of the spleen to haematopoietic maturation has long been established. There is now evidence that the spleen is also influential in the expression of cell-surface markers. Thymocytes from certain strains of newborn mice are capable of undergoing in vitro blastogenesis in response to stimulation from adult, isogeneic spleen cells (58,59) or lymph-node cells.(59) This response, as shown in Table V, has been termed the type 1

TABLE V

TYPES OF
ISOGENEIC LYMPHOCYTE INTERACTIONS (ILI)

Type ILI	Responder cells	Stimulator cells*	^{3}H-Thymidine incorporation at 72 hours		S.I.
			EXPERI-MENTAL	BACK-GROUND	
1	Neonatal thymus	Adult spleen	9,640	361	26.2
1	Neonatal thymus	Adult LN	5,508	661	8.3
2	Adult LN	Adult spleen	2,304	693	3.3
-	Adult LN	Adult LN	961	838	1.1

* Mitomycin-C treated

isogeneic lymphocyte interaction (ILI) and is stimulated by a B-cell surface antigen designated as murine differentiation antigen 1, MDA-1. An analogous, albeit weaker, response exists in adult lymph-node cells to isogeneic adult spleen cells.(60) This phenomenon, known as type 2 ILI, is provoked by another cell surface antigen, MDA-2. The immunological significance of the two B-cell antigens, MDA-1 and

MDA-2, and the respective isogeneic responses which they stimulate is as yet unknown, but recent information suggests that MDA-1 may be essential for B cell interaction with a T cell factor.(61) Specifically, the ability of splenic B cells to respond to a T-cell product, allogeneic effect factor (AEF), by synthesizing IgM and IgG antibodies correlates with expression of the MDAs on the splenic B cell surface. Whether the MDAs are directly involved in interaction with AEF for antibody synthesis or whether they are mediaries in a more complex communication system is unknown. Further definition of the relationship of MDAs to antibody synthesis is currently in progress.

Splenic tissue appears essential to maturation of MDA-2 since the antigen is expressed only on bone-marrow cells which mature in the spleens of irradiated, bone-marrow reconstituted mice.(59) A direct influence of the spleen on the expression of MDA-1 appears unlikely since adult lymph-node cells from neonatally-splenectomized mice are also stimulatory in a type 1 ILI.(60) However, as shown in Table VI, the spleen may be indirectly responsible for maturation

TABLE VI

SPLENIC T CELLS ASSIST MATURATION OF
MDA-1 ON BONE-MARROW CELLS RESIDING IN THE SPLEEN

Spleen tested at:	Irradiated mice reconstituted with BM cells and:	CPM Experi-mental	Back-ground	S.I.	% of control S.I.
5 weeks	Adult spleen (control)	1,003	89	11.2	-
	Nothing	273	115	2.3	20.6
	Splenic T cells	3,385	223	15.0	139.0
	Thymic supportive tissue	1,201	161	7.4	67.8
6 weeks	Adult spleen (control)	12,833	454	28.2	-
	Nothing	783	342	2.2	7.7
	Thymocytes	2,677	384	6.8	24.0

of MDA-1 since splenic, but not thymic, T cells promote earlier maturation of this antigen on B cells.

There is evidence to support the fact that MDA-1 and MDA-2 are closely associated with alloantigens of the I region of the mouse major histocompatibility complex.(59) The I-region-associated (Ia) antigens have been associated with such immunological functions as mixed lymphocyte reactions, GVH reactivity and cell-cell interactions.(62,63) As suggested by the preliminary data on antibody synthesis, the spleen may be actively providing a signal or receptor, in the form of MDAs, for any of these diverse functions. Alternatively, the type I ILI seen in neonatal thymocytes may be essential to establishment of self-recognition and self-tolerance which prevents autoimmunity. Clearly, the bailiwick of splenic function expands as cellular mechanisms are elucidated.

AUTOIMMUNITY

In view of the fact that the spleen has been shown to provide regulatory suppressor function and that autoimmune diseases may in some instances be attributed to a lack of suppression, the occurrence of spontaneous autoimmunity in asplenic mice has been investigated. Studies have thus far shown no positive data to support the concept that absence of the spleen leads to spontaneous induction of autoantibodies. Sera of Dh/+ and +/+ mice from one to 12 months of age were examined for anti-nuclear and thymocytotoxic antibodies as well as for antibodies to native DNA and synthetic double-stranded RNA.(12) Since none of the sera tested positively, absence of the spleen does not, of itself, induce autoreactivity. However, numerous studies indicate that the spleen has important regulatory control of autoimmunity once this condition is established.

Interesting data on immunological regulation by the spleen of antibodies to DNA and RNA have recently been obtained from NZB/NZW F_1 mice. The offspring of New Zealand Black (NZB) and White (NZW) mice (the animal models for studies of autoimmunity, 64), spontaneously produce antibodies to nucleic acids and develop autoimmune disease associated with immune complex glomerulonephritis (65) Neonatal splenectomy of F_1 female mice has been shown to produce, by one month of age, markedly decreased levels of both IgM and IgG antibodies to double-stranded DNA and Poly A, a synthetic single-stranded RNA. For male F_1 mice, the decrease was evident only for IgM antibodies to DNA and Poly A. Data on neonatal thymectomy and/or splenectomy of male and female mice suggest that production of spontaneous antibodies to nucleic acids is strongly dependent upon regulatory controls by the spleen and thymus and is, indeed, a problem of disordered immunological function. Appearance of antibodies to DNA and poly A is asynchronous and removal of the spleen and/or thymus affects the two responses differently, Data suggest that the predominant effect of splenic and thymic ablation in the male is elimination of suppressor cells while, in the female, it appears to be removal of a helper function, Studies into this apparent dichotomy of regulation which results from interaction of suppressor and helper cells (66) should provide significant insight into the mechanisms of autoimmune processes,

TUMOUR IMMUNITY

The defensive role of interferon against invasion of the body by viruses and other parasites is particularly important during the early phases of infection when specific immunity has not yet developed. Evidence indicates that splenic T cells produce interferon, although macrophages and B cells are also capable of interferon production under suitable conditions.(67) Investigations of the role of the spleen in the production of virus-induced interferon following splenectomy have resulted in both positive (68) and negative (69) data. Recent studies have been conducted to test the ability of hereditarily-asplenic mice to produce interferon in response to challenge by Newcastle-disease virus.(70) Data indicate that asplenic mice do have a markedly depressed ability to synthesize interferon. Furthermore, the interferon production can be restored to normal levels by infusion of young spleen cells to neonatal asplenic mice. Thus, the defect appears to be the direct result of asplenia.

Although the role of the spleen in oncogenesis is far from clear, numerous studies have shown that splenic regulation strongly affects immune responses to tumours, However, the role of the spleen remains abstruse since both beneficial and detrimental results have been found in tumour-bearing hosts following splenectomy.(71,72) In one study, splenectomy produced both positive and negative effects on growth depending on the dose of tumour cells injected.(72) The spleen has also been found to regulate lymphocyte trapping in lymph nodes which drain tumour grafts,(73) thus providing another example of splenic control on other immunological domains,

Observation of asplenic and normal mice for more than one year has shown that the two groups do not differ in the sponstaneous incidence of malignant neoplasia nor in their neonatal susceptibility to tumour induction by polyoma virus.(12,74) Asplenic mice also display a normal susceptibility and latency period to Moloney sarcoma virus but develop a larger tumour and require a longer time for tumour regression.(12,74) In contrast, asplenia provides resistance to papilloma induction by gavage with dimethylbenzanthracene.(74) Further experiments with the effects of a splenic transplant on papilloma induction should provide important information to the cellular theory of skin neoplasia,

Evidence does exist that the spleen provides suppressor cell activity in tumour-bearing mice,(75,76) Specifically, recent data describe a soluble factor,(77) elaborated by splenic cells from fibrosarcoma-bearing mice, which inhibits DNA synthesis and DNA polymerase activity of mitogen-stimulated lymphocytes. Therefore, a mechanism for immunosuppression of cellular proliferation originates in splenic cells. Perhaps these data are but the first indications of a vast regulatory system, governed by the spleen, which awaits elucidation.

SUMMARY

Data thus far available unequivocally indicate that the regulatory role of the spleen in immunological processes is extensive. Not only can changes be observed in quantities of immunoglobulins but there are also marked alterations in antigenic stimulation, maturation of T-helper cells for antibody synthesis, mitogenic responses, graft rejection and homing patterns of thymic T cells. Furthermore, the spleen strongly influences the acquisition of at least two B-cell surface antigens and is an essential organ for regulation of the suppressor arm of the immunological network. In view of the data presented in this chapter, the spleen emerges as a lymphoid organ which strongly influences lymphoid cells in other tissues, The mechanisms by which the spleen exerts the various effects that have been described are largely unknown. Soluble splenic factors, distinct splenic cell populations or mediation at both molecular and cellular levels may be involved.

Thorough examination of the data within this chapter adds still another perspective to splenic function: namely, the primary cell which mediates splenic control is the T cell, As seen in Table VII, all responses which have been shown to be altered by asplenia, whether congenital or acquired, involve functional T cells, Although some of these immunological parameters depend also on B cells, the moiety common to all of the observations is the T cell, The principle thus emerges that the spleen is a lymphoid organ which not only affects other lymphoid domains but that it specifically exerts its impact upon distinct subpopulations of T cells, Our understanding of the spleen is advancing from one in which this lymphoid organ is chiefly oriented to immunity mediated by antibody

TABLE VII

SPLENIC EFFECTS UPON OTHER IMMUNOLOGICAL DOMAINS

Response affected	Cell affected
B and T cell cooperation	primarily T
Responses of BM and LN cells to PHA	T
Thymocytes homing to spleen	T
Suppressor cell maturation and/or suppressor factor elaboration	T
MDA maturation	B alone; B through T

reactions to the newer and expanded categorization of the spleen as a dynamic organ which is also strongly influential in cellular immunity. This new perspective of the spleen is strongly supported by much of the recent immunological literature.

REFERENCES

1. BATTISTO, J.R. ET AL. Splenic determination of immunocompetence: influence on other lymphoid organs. Cellular Immunology, 2: 627 (1971).

2. ROMBALL, C.G. and WEIGLE, W.O. Splenic role in the regulation of immune responses. Cellular Immunology, 34: 376 (1977).

3. FABRI, P.J. ET AL. A quarter century with splenectomy, Archives of Surgery, 108: 569 (1974).

4. KEVY, S.V. ET AL. Hereditary splenic hypoplasia. Pediatrics, 42: 752 (1968).

5. WALDMAN, J.D. ET AL. Sepsis and congenital asplenia, Journal of Pediatrics, 90: 555 (1977).

6. BOWDLER, A.J. The role of the spleen and splenectomy in autoimmune hemolytic disease, Seminars in Hematology, 13: 335 (1976).

7. HANCOCK, B.W. ET AL. The immediate effects of splenectomy, radio-therapy and intensive chemotherapy on the immune status of patients with malignant lymphoma. Clinical Oncology, 3: 137 (1977).

8. SEARLE, A.G. Hereditary absence of spleen in the mouse. Nature, 184: 1419 (1959).

9. GREEN, M.C. A defect of the splanchnic mesoderm caused by the mutant gene dominant hemimelia in the mouse. Developmental Biology, 15: 62 (1967).

10. MEIER, H. and HOAG, W.G. Blood proteins and immune response in mice with hereditary absence of spleen. Naturwissenschaften, 49: 329 (1962).

11. LOZZIO, B.B. Hematopoiesis in congenitally asplenic mice. American Journal of Physiology, 222: 290 (1972).

12. FLETCHER, M.P. ET AL. Splenic influence of T cell function: the immunobiology of the inbred hereditarily asplenic mouse. Journal of Immunology, 119: 110 (1977).

13. FIOCCHI, C. Quantitation of immunoglobulins in hereditarily asplenic mice. In: Immuno-aspects of the Spleen, Proceedings of a Conference on Immunodynamics, Cleveland, Ohio, 19-20 May 1976. Amsterdam, Elsevier/North-Holland Biomedical Press, 1976, p. 427.

14. FAHEY, J.L. and BARTH, W.F. The immunoglobulins of mice. 4. Serum immunoglobulin changes following birth. Proceedings of the Society for Experimental Biology and Medicine, 118: 596 (1965).

15. CLARET, I. ET AL. Immunological studies in the post-splenectomy syndrome. Journal of Pediatric Surgery, 10: 59 (1975).

16. KALPAKTSOGLOU, P.K. and GOOD, R.A. Mortality and five classes of mouse immunoglobulins in early splenectomy. International Archives of Allergy and Applied Immunology, 44: 697 (1973).

17. ROZING, J. ET AL. Effects of splenectomy on the humoral immune system. A study in neonatally and adult splenectomized mice. Immunology, 34: 909 (1978).

18. NISHIOKA, K. ET AL. The characteristics, isolation and synthesis of the phagocytosis stimulating peptide tuftsin. Biochemical and Biophysical Research Communications, 47: 172 (1972)

19. CONSTANTOPOULOS, A. ET AL. Defective phagocytosis due to tuftsin deficiency in splenectomized subjects. American Journal of Diseases of Children, 125: 663 (1973).

20. LYNN, H.B. A re-evaluation of splenectomy. Pediatric Annals, 5: 12 (1976).

21. HAAIJMAN, J.J. ET AL. Immunoglobulin-containing cells in different lymphoid organs of the CBA mouse during its lifespan. Immunology, 32: 427 (1977).

22. BATTISTO, J.R. ET AL. Immunoglobulin synthesis in hereditarily spleenless mice. Nature, 222: 1196 (1969).

23. WELLES, W.L. and BATTISTO, J.R. Splenic input on immune capability of lymphoid cells. In: Immuno-aspects of the Spleen, Proceedings of a Conference on Immunodynamics, Cleveland, Ohio, 19-20 May 1976. Amsterdam, Elsevier/North-Holland Biomedical Press, 1976, p. 157.

24. LANDAHL, C.A. ET AL. Studies on the maturation of immune responsiveness in the mouse. II. Role of the spleen. Journal of Immunology, 117: 151 (1976).

25. LOZZIO, B.B. and WARGON, L.B. Immune competence of hereditarily asplenic mice. Immunology, 27: 167 (1974).

26. WARGON, L.B. and LOZZIO, B.B. Antibody response to lipopolysaccharide and sheep erythrocytes by congenitally asplenic mice. IRCS, 2: 1675 (1974).

27. BOREK, F. ET AL. Distribution of sheep red blood cells in tissues of normal and hereditarily spleenless mice. Federation Proceedings, 31: 745 Abs. (1972).

28. MACHADO, E.A. and LOZZIO, B.B. Hyposplenia, asplenia and immunodeficiency. American Journal of Pathology, 85: 515 (1976).

29. BOREK, F. Distribution of sheep red blood cells in tissues of normal and hereditarily spleenless mice. In: Immuno-aspects of the Spleen, Proceedings of a Conference on Immunodynamics, Cleveland, Ohio, 19-20 May 1976. Amsterdam, Elsevier/North-Holland Biomedical Press, 1976, p. 435.

30. NOSSAL, G.J.V. and PIKE, B.L. Differentiation of B lymphocytes from stem cell precursors. Advances in Experimental Medicine and Biology, 29: 11 (1973)

31. CHANANA, A.D. ET AL. Thymus cell migration: [3]HTdR-labelled and theta-positive cells in peripheral lymphoid tissues of newborn mice. Advances in Experimental Medicine and Biology, 29: 79 (1973).

32. SPEAR, P.G. ET AL. Characterization of splenic lymphoid cells in fetal and newborn mice. Journal of Experimental Medicine, 138: 557 (1973).

33. SPEAR, P.G. ET AL. Maturation of the humoral immune response in mice. Journal of Experimental Medicine, 139: 249 (1974).

34. RABINOWITZ, S.G. Measurement and comparison of the proliferative and antibody responses of neonatal, immature and adult murine spleen cells to T-dependent and T-independent antigens. Cellular Immunology, 21: 201 (1976).

35. BUCSI, R.A. ET AL. Splenic replenishment of synergistic ability to bone marrow and thymus cells of neonatally splenectomized CBA mice. Journal of Experimental Medicine, 136: 761 (1972).

36. SHILLCOCK, J.A. ET AL. Effect of neonatal splenectomy on cellular cooperation in IgM and IgG synthesis. Federation Proceedings, 32: 966 Abs. (1973).

37. WARGON, L.B. ET AL. Alteration of bone marrow-thymus cell synergism in hereditarily asplenic and adult-splenectomized mice. Proceedings of the Society for Experimental Biology and Medicine, 148: 925 (1975).

38. RUSSELL, J.L. and GOLUB, E.S. Functional development of the interacting cells in the immune response. III. Role of the neonatal spleen. European Journal of Immunology, 7: 305 (1977).

39. AUERBACH, R. Ontogeny of immune responsiveness in the absence of the spleen: a review. Developmental and Comparative Immunology, 2: 219 (1978).

40. BUCSI, R.A. and BATTISTO, J.R. Splenic imprint on lymphoid cells. Experimental Hematology, 1: 72 Abs. (1973).

41. SY, M. ET AL. A splenic requirement for the generation of suppressor T cells. Journal of Immunology, 119: 2095 (1977).

42. GARCIA-GIRALT, E. ET AL. Prevention of graft versus host reaction by incubation of lymphoid cells with a splenic extract. Cell Tissue Kinetics, 6: 567 (1973).

43. STREILEIN, J.W. and WIESNER, J. Influence of splenectomy on first set rejection reactions of C57BL/6 females to male skin isografts. Journal of Experimental Medicine, 146: 809 (1977). (1977).

44. COONS, T.A. and GOLDBERG, E.H. Rejection of male skin grafts by splenectomized female mice. Science, 200: 320 (1978).

45. GERSHON, R.K. A disquisition on suppressor T cells. Transplantation Reviews, 26: 170 (1975).

46. TADA, T. ET AL. Properties of primed suppressor T cells and their products. Transplantation Reviews, 26: 106 (1975).

47. PIERCE, C.W. and KAPP, J.A. Regulation of immune responses by suppressor T cells. Contemporary Topics in Immunobiology, 5: 91 (1976).

48. GERSHON, R.K. T-cell control of antibody production. Contemporary Topics in Immunobiology, 3: 1 (1974).

49. BASTEN, A. ET AL. Cell-to-cell interaction in the immune response. X. T-cell dependent suppression in tolerant mice. Journal of Experimental Medicine, 140: 199 (1974).

50. MOSIER, D.E. and JOHNSON, B.M. Ontogeny of mouse lymphocyte function. II. Development of the ability to product antibody is modulated by T lymphocytes. Journal of Experimental Medicine, 141: 216 (1975).

51. BASSETT, M. ET AL. Suppression of stimulation in mixed leukocyte culture by newborn splenic lymphocytes in the mouse. Journal of Immunology, 119: 1855 (1977).

52. ARGYRIS, B.F. Suppressor activity in the spleen of neonatal mice. Cellular Immunology, 36: 354 (1978).

53. McCULLAGH, P. Suppressor cells in homograft tolerant rats. Australian Journal of Experimental Biology and Medical Science, 53: 431 (1976).

54. RIEGER, M. and HILGERT, I. The involvement of a suppressor mechanism in neonatally induced allograft tolerance in mice. Journal of Immunogenetics, 4: 61 (1977).

55. ZEMBALA, M. ET AL. Contact sensitivity to picryl chloride: the occurrence of B suppressor cells in the lymph nodes and spleen of immunized mice. Cellular Immunology, 25: 266 (1976).

56. KINCHNER, J. ET AL. Splenic suppressor macrophages induced in mice by injection of Corynebacterium parvum. Journal of Immunology, 115: 1212 (1975).

57. GERSHON, R.K. ET AL. Immuno-regulatory role of spleen localizing thymocytes. Journal of Immunology, 112: 546 (1974).

58. HOWE, M.L. ET AL. Isogeneic lymphocyte interaction: recognition of self-antigens by cells of the neonatal thymus. Proceedings of the National Academy of Sciences, 67: 613 (1970).

59. FINKE, J.H. and BATTISTO, J.R. Maturational requirements of murine differentiation antigens 1 and 2 that are operative in isogeneic lymphocyte interactions. In: Immuno-aspects of the Spleen, Proceedings of a Conference on Immunodynamics, Cleveland, Ohio, 19-20 May 1976. Amsterdam, Elsevier/North Holland Biomedical Press, 1976, p. 89.

60. PONZIO, N.M. ET AL. Adult murine lymph node cells respond blastogenically to a new differentiation antigen on isologous and autologous B lymphocytes. Journal of Immunology, 114: 971 (1975).

61. BATTISTO, J.R. ET AL. Correlation of B cell acquisition of differentiation antigens with capacity to interact with allogeneic effect factor (AEF). In: VIII International Congress of the Reticuloendothelial Society, Jerusalem, Israel, 18-23 June 1978. In press.

62. SCREFFLER, D.C. and DAVID, C. The H-2 major histocompatibility complex and the I immune response region: genetic variation, function and organization. Advances in Immunology, 20: 125 (1975).

63. NIEDERHUBER, J.E. and FRELINGER, J.A. Expression of Ia antigens on T and B cells and their relationship to immune-response functions. Transplantation Reviews, 30: 101 (1976).

64. ROUBINIAN, J.R. ET AL. Immunological regulation of spontaneous antibodies to DNA and RNA. III. Early effects of neonatal thymectomy and splenectomy. Immunology, 33: 399 (1977).

65. HOWIE, J.B. and HELYER, B.J. The immunology and pathology of NZB mice. Advances in Immunology, 9: 215 (1968).

66. TALAL, N. Disordered immunologic regulation and autoimmunity. Transplantation Reviews, 31: 240 (1976).

67. TSUKUI, K. Influenza virus-induced interferon production in mouse spleen cell culture: T cells as the main producer. Cellular Immunology, 32: 243 (1977).

68. FRUITSTONE, M.J. ET AL. Role of the spleen in interferon production in mice. Proceedings of the Society for Experimental Biology and Medicine, 122: 1008 (1966)

69. BORECKY, L. and LACKOVIC, V. The cellular background of interferon production in vivo. Acta Virologica, 11: 150 (1967).

70. LAIR, S.V. ET AL. Induction of interferon in hereditarily asplenic mice with and without a neonatal spleen cell transplant. Proceedings of the Society for Experimental Biology and Medicine, 146: 475 (1974).

71. MILAS, L. and MUJAGIC, H. The effect of splenectomy on fibrosarcoma "metastases" in lungs of mice. International Journal of Cancer, 11: 186 (1973).

72. NORDLUND, J.J. and GERSHON, R.K. Splenic regulation of the clinical appearance of small tumors. Journal of Immunology, 114: 1486 (1975).

73. FIGHTLIN, R.S. ET AL. Splenic regulation of lymphocyte trapping in lymph nodes draining tumor grafts. Journal of Immunology, 115: 345 (1975).

74. FLETCHER, M.P. ET AL. Oncogenesis in congenitally asplenic mice. In: Immuno-aspects of the Spleen, Proceedings of a Conference on Immunodynamics, Cleveland, Ohio, 19-20 May 1976. Amsterdam, Elsevier/North-Holland Biomedical Press, 1976, p. 377.

75. POPE, B.L. ET AL. Suppressor cells in the spleens of tumour-bearing mice. Journal of Immunology, 116: 1342 (1976).

76. KIRCHNER, H. ET AL. Evidence of suppressor cell activity in spleens of mice bearing primary tumors induced by Moloney sarcoma virus. Journal of Experimental Medicine, 139: 1473 (1974).

77. FARRAR, W.L. and ELGERT, K.D. Suppressor cell activity in tumor-bearing mice. II. Inhibition of DNA synthesis and DNA polymerases by TBH splenic suppressor cells. Journal of Immunology, 120: 1354 (1978).

DISCUSSION

MITCHELL: Has anyone yet determined the number of cells of various Ly phenotypes in the lymph nodes of asplenic mice to determine whether there is any obligatory activity of the spleen in the processing of thymus cell migrants?

BATTISTO: To my knowledge there has, as yet, been no assessment of the Ly antigens on lymph nodal cells of asplenic mice. Such an examination might prove profitable.

MITCHELL: There is a report that the thymocytes of asplenic mice respond to LPS, an observation which implies that the thymus of asplenics is very different from the thymus of intact mice and may be invaded by B-cell elements. Have you been able to confirm this finding?

BATTISTO: It is, of course the provocative observation made by Drs Russell and Golub to which you refer. In the recent paper in the Journal of Immunology by Fletcher, Ikeda and Gershvin there was mention of an inability to repeat this observation. Thus, whether B cells exist in the thymi of Dh/+ mice is still open to question.

SPRENT: You mentioned that homing to the spleen by thymus cells from asplenic mice is reduced. Where do these cells go?

BATTISTO: There are no reports upon where these thymus cells go. The report by Drs Russell and Golub simply shows that, at 160 hours, about 40% of those thymic cells derived from "spleened" mice can be found in the spleen of an x-irradiated recipient whereas half that amount of thymic cells coming from asplenic mice are found to become localized in the spleen.

FORD: Are there any data on the size of the recirculating lymphocyte pool in asplenic mice? If it were reduced, then the decreased proportion of thymus cells which were found to localize in the spleen may be explained by a reduction in the small minority of recirculating lymphocytes in the thymus medulla found by Rannie and Donald.

BATTISTO: There are no data of which I am aware on the size of the recirculating lymphocyte pool in asplenic mice. This obviously should be looked at by someone competent in this particular area.

HUMPHREY: When you have shown that the failure of T-B cooperation on neonatally splenectomized mice can be restored by implantation of a syngeneic spleen, can this be done with a spleen irradiated with - say - 100 rads, sufficient to destroy its contained lymphocytes (or with spleen from a recently irradiated mouse)? This might indicate whether the effect was mediated by the "structural" cells rather than the migratory cells of the spleen.

BATTISTO: We had not utilized x-irradiated spleens in the restoration experiments. I agree that the use of such spleens might differentiate which cells are responsible for the restitution. We should go back to look at this aspect once more.

The Spleen and Splenic Lymphocytes in Immunity

G.F. Mitchell, Laboratory of Immunoparasitology,
The Walter & Eliza Hall Institute of Medical Research,
Royal Melbourne Hospital P.O., Victoria, Australia.

The spleen is an organ which is obviously involved in immune responses to blood-borne antigen yet its precise role in immune induction and, in particular, immune regulation, is far from clear. Several important questions on the immunological status of the spleen relate to whether the splenic lymphocyte population differs in any significant way from the lymphocyte population in the secondary lymphoid organs in general. Is the splenic cell population not able to mediate any particular immune response in situ, in vitro or on adoptive transfer? Is the spleen enriched for, or deficient in, any particular T or B cell subpopulation? In ontogeny, are there differences in the time course of development of major or minor lymphocyte subpopulations? Is the spleen basically equivalent to any visceral lymph node but with active haemopoiesis (in rodents) and red-cell destruction superimposed? In this discussion, some of these questions will be addressed, the emphasis being on studies in laboratory mice and rats.

MONONUCLEAR CELLS IN SPLEEN

One feature of the spleen, often carelessly disregarded, is that not every mononuclear cell in the spleen is a lymphocyte. The proportion of monocytes and normoblasts can be expected to be high in the spleen particularly in situations of splenomegaly such as in hemosporidial protozoan infections. Agranular mononuclear cells in spleen may be T cells of various subpopulations (1) or pre B cells, virgin B cells and memory B cells at various stages of differentiation, (2,3) but they may also be monocytes, macrophages, normoblasts or stem cells.(4). Thus it is foolhardy to talk of the surface markers or immunological activities of splenic lymphocytes (in adoptive transfer or in vitro experiments) using uncharacterized dispersed cell suspensions of spleen.   Dilution effects of splenic lymphocytes with normoblasts in malarious mice can account for the reduced capacity of spleen cells to generate a PFC response in vitro.(5) Another common mistake is to assume that lymphoid cells in the spleen of hypothymic (nude) mice are principally B cells since many other mononuclear cells are present including large numbers of a short-lived, non-recirculating, Ig^-, weakly θ^+ (Thy^+) cell population.(6)

ONTOGENY OF SPLENIC T AND B LYMPHOCYTES

Using dispersed cell suspensions and a detection system of radiolabelled polyspecific anti Ig sera and autoradiography, B cells are readily detected in the mouse spleen at birth but are absent in lymph nodes. The number of B cells in mesenteric lymph node increases rapidly during the first week of life whereas the major increase in splenic B cells commences, along with other erythromyelopoietic organs, prior to birth.(7) This situation contrasts with that for T cells in which the mesenteric lymph node contains many more T cells than does the spleen in the first few days of life.(8) At least very early in life, T cell migrants appear not to home to the spleen and all available evidence in the mouse indicates that in ontogeny, the spleen contains B cells (i.e. Ig^- cells) before it contains T cells (i.e. Ig^+ $Thy\ 1^+$ cells) and that the lymph nodes contain T cells before they contain B cells It has been proposed that, later in life, T cell migrants go to spleen preferentially (9-12) where interaction with antigens occurs and progeny of the antigen-driven expansion then join the recirculating pool.(13) At one week of age, all $Thy\ 1^+$ spleen cells are $Ly-123^+$ and as the mouse matures the proportion of $Ly-123^+$ T cells declines in favour of $Ly-1^+$ and $Ly-23^+$ cells; presumably $Ly-123^+$ cells are "young" thymus migrants since this subclass in spleen is reduced by adult thymectomy.(1)

DIFFERENCES BETWEEN SPLENIC AND LYMPH NODE
T AND B LYMPHOCYTE POPULATIONS

In a scheme of lymphopoiesis, there seems to be some justification in placing the spleen between unequivocally primary lymphoid organs such as the thymus (and bone marrow in early life) and unequivocally secondary lymphoid organs such as lymph nodes.(14) With respect to B cells, the spleen differs from lymph nodes in being a source of pre B cells (15) and it is enriched for virgin B cells; (3) in keeping with this, Goding et al. (16) have shown that, for any given age, cells bearing both μ and δ Ig markers represent a smaller fraction of the total B cells in spleen that in lymph node. Moreover, in adult life, the frequency of specific hapten-reactive B cell precursors of IgM antibody-secreting cells in spleen is at least 4x that in mesenteric or subcutaneous lymph nodes (Pike, B.L., personal communication). It is known that the spleen is a major site of IgM synthesis (17) and that splenectomy reduces levels of IgM in the circulation.(18)

Concerning T cells, and as mentioned above, the spleen is probably involved (at least in adult life) in the "processing" of early thymus migrants to potentially long-lived recirculating cells although it must be pointed out that much of the experimentation on this point involves injection of a dispersed cell suspension of thymus to recipients; such experiments are far from physiologic. Relative to lymph nodes, the spleen of antigen-injected mice may contain greater T cell-dependent immunosuppressive activity on transfer to irradiated recipients.(12,19-21) However, to the author's knowledge, this difference has not yet been correlated with differences in T cell subpopulations in spleen versus lymph nodes as assessed by lymphocyte surface markers (e.g. Ly and IJ).

Induction of immune responses to cellular (e.g. MHC and SRBC) antigens in the spleen rather than lymph nodes of rodents leads to "enhancing" antibodies (22,23) and defective cell-mediated immunity. (24,25) Kaplan and Streilen (24) have proposed that one reason for "immunological privilege" of transplants in locations such as the anterior chamber of the eye is that antigen gains direct access to blood (and thus spleen) rather than lymph (and thus lymph nodes) thereby promoting production of antibodies (perhaps of certain isotypes) at the expense of aggressive cell-mediated immunity.

An important unresolved question is whether certain Ig isotypes are preferentially induced in spleen versus lymph nodes. It is known that production of IgE (26-30) and IgA (31-33) is essentially extrasplenic. However, what is not known is whether the different IgG isotypes (subclasses) which have very different biological functions, are induced differentially in spleen versus lymph nodes. This will be discussed further in the next section.

EFFECTS OF SPLENECTOMY

The classical means of searching for any unique or dominant function(s) of an organ is to remove that organ and to study the consequences. Despite several interesting observations (see Battisto, J.R., this volume), analyses of the immunocompetence of asplenic and splenectomized mice have not yet been as informative as one would have hoped. It is safe to say that there has not yet been a demonstration of a unique activity of spleen which can not be taken over by other organs. Perhaps this is to be expected of a secondary lymphoid organ within the circulating pool of lymphocytes (one particularly informative study would be to examine the Ly markers on lymph-node lymphocytes from asplenic mice). Regarding circulating Ig levels, asplenic mice are reported to have reduced levels of IgM and IgG_2 whereas IgG_1 levels are normal.[34] A T cell-dependent IgG_1 hypergammaglobulinaemia is very prominent in mice chronically infected with metazoan parasites [35-38] but there is no evidence as yet that non-splenic lymphoid sites are the principal sites of IgG_1 production. Unexpectedly, the thymocytes from asplenic mice have been found to respond to LPS [39] and, more predictably,[40] the bone marrow of asplenics behaves like spleen in adoptive immune responses in irradiated mice.[41] In all studies on splenectomy in conventional animal colonies, one must contend with Hemobartonella (rats) and Eperythrozoon (mice) infections which may flare up and complicate immunological and haematological analyses.

The splenic vein joins the portal system. It is known that after intravenous injection of antigen, the bulk of the antibody formed comes from cells located in the spleen [42,43] and that this antibody enters the portal system via the splenic vein.[42,44] This anatomical plumbing suggests that opsonizing antibodies (e.g. IgM and IgG_{2a} in the mouse), derived from the spleen and with specificity for antigens coming from the gut, are important in the clearance and destruction of antigen in the Kupffer cells of the liver.[44-46] Such antibody-mediated elimination of gut-derived antigen present in the portal system may be instrumental in preventing exaggerated sensitization not only to food antigens and intestinal microorganisms,[44] but also to antigens of metazoan and protozoan parasites resident in the gut. It would be of particular interest to determine the effect of anastomosis of the splenic vein to the vena cava on the level of sensitization to gut-derived antigens and immune responses to gut parasites. If the notion has validity, the effects of splenectomy may not be as dramatic as surgical deviation of the splenic venous outflow since removal of the spleen will remove not only the source of opsonizing antibodies but also the major site where immune responses to blood-borne antigen occur.

The present state of knowledge enables one to make a qualified case for the idea that the spleen, although containing T cells fully able to transfer cell-mediated immunities (e.g. 47,48) and to respond by proliferation,[49] is an organ which in situ tends to become involved in responses to blood-borne antigens, in relatively high doses, by production of high titered IgM and IgG antibodies. These splenic antibody responses may not be induced at the expense of T cell sensitization; rather they may be the cause of reduced cellular sensitivity as expressed in DTH and other cell-mediated immune reactions.

REFERENCES

1. CANTOR, H. and BOYSE, E.A. Development and function of subclasses of T cells. Journal of the Reticuloendothelial Society, 17: 115 (1975).

2. NOSSAL, G.J.V. ET AL. Current problem areas in the study of B lymphocyte differentiation. Immunological Reviews, 37: 187 (1977).

3. STROBER, S. Immune function, cell surface characteristics and maturation of B cell subpopulations. Tranplantation Reviews, 24: 84 (1975)

4. METCALF, D. and MOORE, M.A.S. Haemopoietic Cells. North-Holland, Amsterdam (1971).

5. FREEMAN, R.R. Malaria Infections in Laboratory Mice. Ph.D. thesis, Australian National University, Canberra, Australia (1978).

6. ROELANTS, G. ET AL. Five types of lymphocytes ($Ig^- \theta^-$, $Ig^- \theta^{+weak}$, $Ig^- \theta^{+strong}$, $Ig^+ \theta^-$, and $Ig^+ \theta^+$) characterized by double immunofluorescence and electrophoretic mobility. Organ distribution in normal and nude mice. European Journal of Immunology, 5: 127 (1975).

7. NOSSAL, G.J.V. and PIKE, B.L. Studies on the differentiation of B lymphocytes in the mouse. Immunology, 25: 33 (1973).

8. WEISSMAN, I.L. Development and distribution of immunoglobulin-bearing cells in mice. Transplantation Reviews, 24: 159 (1975).

9. CANTOR, H. Two stages in development of lymphocytes. In: Cell interactions, (L.G. Silvestri, ed.) North-Holland, Amsterdam, p. 172 (1972).

10. BACH, J.-F. and DARDENNE, M. Antigen recognition by T lymphocytes. III. Evidence for two populations of thymus-dependent rosette-forming cells in spleen and lymph nodes. Cellular Immunology, 6: 394 (1973).

11. OLSSON, L. and CLAESSON, M.H. Studies on subpopulations of theta-bearing lymphoid cells. Nature, 244: 50 (1973)

12. GERSHON, R.K. T cell control of antibody production. Contemporary Topics in Immunobiology, 3: 1 (1973).

13. MILLER, J.F.A.P. and MITCHELL, G.F. Interactions between two distinct cell lineages in an immune response. In: Lymphatic Tissue and Germinal Centers in Immune Response. (L. Fiore-Donati and M.G. Hanna, Jr., eds) Plenum Press, New York, p.455 (1969).

14. MITCHELL, G.F. Observations and speculations on the influence of T cells in the cellular events of induction of antibody formation and tolerance in vivo. In: The Lymphocyte: Structure and Function (J.J. Marchalonis, ed.) Marcel Dekker, New York, p. 227 (1977).

15. LAFLEUR, L. ET AL. Restriction of specificity in the precursor of bone marrow-associated lymphocytes. Journal of Experimental Medicine, 137: 954 (1973).

16. GODING, J.W. ET AL. Genetics, cellular expression and function of IgD and IgM receptors. Immunological Reviews, 37: 152 (1977

17. HAAJMAN, J.J. ET AL. Immunoglobulin-containing cells in different lymphoid organs of the CBA mouse during its life-span. Immunology, 32: 427 (1977).

18. ROZING, J. ET AL. B-lymphocyte differentiation in lethally irradiated and reconstituted mice. III. The influence of splenectomy on the recovery of the B-cell population. Cellular Immunology, 31: 340 (1977).

19. BASTEN, A. ET AL. T cell-dependent suppression of an anti-hapten antibody response. Transplantation Reviews, 26: 130 (1975).

20. YAMAMOTO, H. ET AL. Regulatory functions of hapten-reactive helper and suppressor T lymphocytes. I. Detection and characterization of hapten-reactive suppressor T-cell activity in mice immunized with hapten-isologous protein conjugate. Journal of Experimental Medicine, 146: 74 (1977).

21. MOORHEAD, J.W. Subpopulations of mouse T lymphocytes. II. Suppression of graft-versus-host reactions by naturally proliferating splenic T cells. European Journal of Immunology, 8: 163 (1978).

22. FERRER, J.F. Role of the spleen in passive immunological enhancement. Transplantation, 6: 167 (1968).

23. ENOMOTO, K. and LUCAS, Z.J. Immunological enhancement of renal allografts in the rat. II. Role of the spleen. Transplantation, 15: 8 (1973).

24. KAPLAN, H.J. and STREILEN, J.W. Do immunologically privileged sites require a functioning spleen? Nature, 251: 553 (1974).

25. LAGRANGE, P.H. ET AL. Influence of dose and route of antigen injection on the immunological induction of T cells. Journal of Experimental Medicine, 139: 528 (1974).

26. MAYRHOFER, G. Sites of synthesis and localization of IgE in rats infested with Nippostrongylus brasiliensis. In: Immunology of the Gut, (R. Porter & J. Knight, eds.) Elsevier, Amsterdam, p. 155 (1977).

27. TADA, T. and ISHIZAKA, K. Distribution of γE-forming cells in lymphoid tissues of the human and monkey. Journal of Immunology, 104: 377 (1970).

28. MITCHELL, G.F. and CLARKE, A.E. The allergenicity of concanavalin A in mice. International Archives of Allergy and Applied Immunology (in press).

29. OGILVIE, B.M. and JONES, V.E. Reaginic antibodies and immunity to Nippostrongylus brasiliensis in the rat. I. The effect of thymectomy, neonatal infections and splenectomy. Parasitology, 57: 335 (1967).

30. BARNETT, J.B. and WUST, C.J. Levels of hemocytotropic antibody in hereditary asplenic, splenectomized and normal mice. International Archives of Allergy and Applied Immunology, 56: 558 (1978).

31. CRAIG, S.W. and CEBRA, J.J. Peyers patches: an enriched source of precursors for IgA-producing immunocytes in rabbits. Journal of Experimental Medicine, 134: 188 (1971).

32. GUY-GRAND, D. ET AL. The gut-associated lymphoid system: nature and properties of the large dividing cells. European Journal of Immunology, 4: 435 (1974).

33. WILLIAMS, A.F. and GOWANS, J.L. The presence of IgA on the surface of rat thoracic duct cells which contain internal IgA. Journal of Experimental Medicine, 141: 335 (1975).

34. WELLES, W.L. and BATTISTO, J.R. Splenic input on immune capability of lymphoid cells. In: Immuno-aspects of the Spleen, (J.R. Battisto & J.W. Streilen, eds.) Elsevier, Amsterdam, p. 157 (1976).

35. CRANDALL, R.B. ET AL. Heligomosoides polygyrus (= Nematospiroides dubius): humoral and intestinal immunologic responses to infection in mice. Experimental Parasitology, 35: 275 (1974).

36. SHER, A. ET AL. Schistosoma mansoni: kinetics and class specificity of hypergammaglobulinaemia induced during murine infection. Experimental Parasitology, 41: 415 (1977).

37. MITCHELL, G.F. ET AL. Studies on immune responses to larval cestodes in mice. Immunoglobulins associated with the larvae of Mesocestoides corti. Australian Journal of Experimental Biology and Medical Sciences, 55: 187 (1977).

38. PROWSE, S.J. ET AL. Immunity to Nematospiroides dubius: cell and immunoglobulin changes associated with the onset of immunity in mice. Australian Journal of Experimental Biology and Medical Sciences, 56: 237 (1978).

39. RUSSEL, J.L. and GOLUB, E.S. Altered T cell function in asplenic mice. In: Immuno-aspects of the Spleen, (J.R. Battisto & J.W. Streilen, eds.) Elsevier, Amsterdam, p. 185 (1976).

40. BENNER, R. and VAN OUDENAREN, A. Antibody formation in mouse bone marrow. IV. The influence of splenectomy on the bone marrow plaque-forming cell response to sheep red blood cells. Cellular Immunology, 19:167 (1975).

41. AUERBACH, R. ET AL. The role of the spleen in the development of immune competence. In: Immuno-aspects of the Spleen, (J.R. Battisto & J.W. Streilen, eds.) Elsevier, Amsterdam, p. 3 (1976).

42. ROWLEY, D.A. The effect of splenectomy on the formation of circulating antibody in the adult male albino rat. Journal of Immunology, 64: 289 (1950).

43. ROMBALL, C.G. and WEIGLE, W.O. Splenic role in the regulation of immune responses. Cellular Immunology, 34: 376 (1977).

44. THOMAS, H.C. and VAEZ-ZADEH, F. A homeostatic mechanism for the removal of antigen from the portal circulation. Immunology, 26: 375 (1974).

45. SLJIVIC, V.S. and WARR, G.W. Activity of the reticuloendothelial system and the antibody response. II. Effect of stilboestrol on the immune response to sheep erythrocytes in the mouse. British Journal of Experimental Pathology, 54: 69 (1973).

46. BRADFIELD, J.W.B. ET AL. The mechanism of the adjuvant effect of dextran sulphate. Immunology, 26: 383 (1974).

93

47. KETTMAN, J. Delayed hypersensitivity: Is the same population of thymus derived cells responsible for cellular immunity reactions and the carrier effect? Immunological Communications, 1: 289 (1972).

48. MILLER, J.F.A.P. and OSOBA, D. Current concepts of the immunological function of the thymus. Physiological Reviews, 47: 437 (1967).

49. DAVIES, A.J.S. The thymus and the cellular basis of immunity. Transplantation Reviews, 1: 43 (1969).

DISCUSSION

SPRENT: As an alternative to a porto-caval shunt, splenectomy followed by subcutaneous grafting of spleens might suffice.

MITCHELL: Yes, provided adequate vascular connexions were established as quickly as possible. Perhaps splenectomy at birth followed by multiple subcutaneous and kidney capsule splenic fragment implantation would be a useful approach.

FRANK: I am embarrassed to note that I do not know the role of opsonin in removal of antigenic materials from the portal vein. Can you clarify this question for me?

MITCHELL: I am also embarrassed to note that I can't clarify this question for you! The accessibility, nature and fate of antigens derived from intestinally-located metazoan and protozoan parasites, for example, are totally unknown but it is safe to predict that as immunologists become increasingly involved in the events of immune induction to "real-life" antigens there will be a flurry of activity in this important research area. Moreover, I think we can also expect to see several very precise analyses published soon on the biological functions (including opsonization) of Ig isotypes in various animals and the circumstances and sites of preferential induction of antibodies of various Ig isotypes. This information is of course obligatory in the rational design of vaccination strategies aimed at inducing desired immune responses at the expense of, or rather than, responses previously determined to be undesirable.

HUMPHREY: Could the route of absorption of radiolabelled potential antigens from the gut be monitored by cannulation of the portal vein and the mesenteric lymph drainage?

MITCHELL: This should certainly be informative although precise measurements would be difficult because, presumably, only a tributary of the portal system would support a long-term indwelling cannula and one could only hope to sample a fraction of the afferent lymph going to the mesenteric lymph node. I am not at all familiar with the literature on access of antigen from the gut via portal versus lymphatic systems and perhaps the information is contained in publications such as the recent Ciba Foundation Symposium on "Immunology of the Gut".

HUMPHREY: When you postulate that antigenic materials are absorbed from the gut via the portal vein are you certain that this is the normal route? I think that it is unsettled whether/which antigens absorbed across intestinal epithelium go into the mesenteric lymphatics rather than the venous drainage. The answer would affect the validity of your hypothesis.

MITCHELL: I have to confess total ignorance and have no idea on what antigenic properties and microenvironmental factors are responsible for access of antigen to bloodstream versus lymphatics. I had assumed that some antigens are absorbed from the gut and gain access to the portal system and that an important function of Kupffer cells plus opsonizing antibodies from the spleen is to restrict access of gut-derived antigens to organized lymphoid tissues. I suspect that this is an area in which hard data are lacking; the idea, which is certainly not my own, is intriguing however.

BATTISTO: You have indicated that DTH cannot be engendered when you observe a great increase in antibody (perhaps IgG_1 antibody). Is this attributable to antigen-masking by the antibody such that antigen is never experienced (as with the Rho-Gam situation) or is the T cell that is involved in DTH diverted to the task of helper for antibody synthesis?

MITCHELL: The simplest explanation is certainly that antibodies reduce the ability of antigen (located at the site of deposition for elicitation of DTH) to ultimately stimulate T cells via whatever presentation mechanisms are involved.

SPRENT: Perhaps splenectomy removes "suppressor" cells which normally limit the production of enhancing antibody?

MORRIS: This might be so, but it is perhaps stretching speculation on mechanism to the limit! I think at the moment we would explain the observations on the basis of a depression of the humoral response to the allogeneic graft, for the humoral response plays a major role in rejection of the renal allograft in the rat. This then leaves us with the parental strain allograft, where rejection still occurs despite the depression of the humoral response, and in this situation it may be that the avidity of antigen reactive T cells for the homozygous parental strain kidney is such that the cellular immune response can alone reject the kidney. Whatever the explanation, it does stress the varying immuno-regulatory role of the spleen, and the transplant model provides a model which has some relevance to the role of the spleen in parasitic infections.

COHEN: Dr Mitchell, could you say more about the parasites causing the IgG_1 increase and its course in relation to infection?

MITCHELL: IgG_1 hypergammaglobulinaemia, involving levels of up to 50 mgm per ml and a much shortened half-life, is seen in mice infected with the larval cestode Mesocestoides conti, the intestinal nematode Nematospiroides dubius, and the tramtode, Schistosoma mansoni. The increase is T-cell dependent in that it is absent in infected nude mice, the amount of anti-parasite antibody contained is not known, and may be seen as early as three weeks of infection. Its candidacy as a "blocking antibody" is attractive but not proven and seems to be a function of chronicity involving persistent and strong antigenic stimulation. We are currently trying to simulate the effect by multiple injections of "strong" antigens (leading to increased IgG levels) versus multiple injections of "weak" antigens (leading to macroglobulinaemia).

MORRIS: Can I briefly switch the discussion to transplantation immunology and mention some work done by John Fabre in Oxford concerning the role of the spleen in kidney allograft rejection in the rat? He found that splenectomy in certain strain combinations performed from six weeks before to three or four days after renal transplantation resulted in prolonged graft survival and suppression of the cytotoxic antibody response to donor strain HC antigens. This effect was marked in the F1 donor situation but where a parental donor was used minimal or no increased graft survival was obtained although the antibody response was still suppressed. Restoration of the response in splenectomized rats could not be obtained with normal syngeneic spleen lymphocytes (10^9) but could be with sensitized spleen lymphocytes. Finally, splenectomy does not interfere with passive enhancement of renal allografts in the rat. Thus in this model of a vascularized organ graft we find a rather different effect of splenectomy to some of the systems that have been discussed today. Can you fit these findings into your hypothesis?

MITCHELL: There is a dilemma here since I know of three references which implicate the spleen as being a site of production of "enhancing" antibodies and there is a recent paper reporting that rejection of male skin in a particular mouse strain only occurs in splenectomized females. Perhaps in the situation (strain combinations?) you have cited, the intact spleen is involved in high titered cytotoxic antibody responses and, from recent results of Blanden et al., the expression of parental strain antigens and cells from certain F_1 hybrids may be much less than the 50% expression one might expect. Thus net cytotoxic antibody production may exceed net enhancing antibody production in the spleen in your experiments and the difference between F_1 and parental kidney donors may relate to antigen dose effects in the target organ.

Immunosuppression in Trypanosomiasis and Malaria

Christine E. Clayton, Division of Immunology,
National Institute of Medical Research, London, England.

Immunosuppression is a frequent accompaniment to disease, and the type and severity of the lesions in the immune response vary as greatly as the causative agent or disease process.(1-4) However, similar defects in immunity are sometimes found in diseases of widely differing etiology. In comparison with other infections, the salivarian trypanosomes affect very many aspects of the immune response, so results obtained in studying this disease are useful indicators to the variety of mechanisms of suppression which may be operating in others. In this paper I shall concentrate on the immunosuppression trypanosomes cause, with some reference to that induced by plasmodia. Both these parasites invade the blood of the host and the resulting immunosuppressions show many common characteristics.

IMMUNOSUPPRESSION IN MAN

The importance of secondary infection in trypanosomiasis has been known for nearly a century. Mott (5) noted in 1906 that 80% of his Ugandan sleeping sickness patients died with streptococcal infection, and diplo-streptococci were so commonly found that Castellani thought initially that they actually caused the disease. Much more recently, Greenwood et al. (6,7) found that the humoral antibody response to some antigens was decreased in children with trypanosomiasis or malaria, the former also inhibiting the response to skin testing.

Multiple infection is the rule in the tropics, and is a particularly severe problem in young children experiencing the first onslaught of the many parasites to which they are exposed. Immunosuppression is of clinical importance because infection with one parasite can increase susceptibility to others.

PATHOLOGY OF THE IMMUNE SYSTEM IN MAN AND PRIMATES

Most of the structural and histological changes in the human immune system during trypanosomiasis were described in the first decade of this century (5,8) by Mott and by Greig and Gray, who noted enlarged lymph glands containing many plasma cells and the mulberry-like "Mott" cells. They saw reticuloendothelial proliferation, a leucocytosis in the circulation, now known to be primarily due to increased lymphocytes and monocytes, and anaemia, with extensive erythrophagocytosis by the reticular cells of the lymph nodes. The only major additions to this description are splenomegaly and the observation of increased levels of serum immunoglobulins which become so elevated as to be a diagnostic feature.(9) The pathology and histology of malaria and trypanosomiasis have been summarized by Edington and Gilles.(10)

EXPERIMENTAL SYSTEMS

Common laboratory models of malaria and trypanosomiasis are summarized in Table I. Most experiments are done with rodents, in which the diseases are often more severe than in man and domestic animals (examples shown for comparison). Although many features of the immunosuppression remain similar, the degree varies - for instance, it is too slight in cattle to impair vaccination.(11) The clinical or veterinary importance of immunosuppression will therefore vary with the host-parasite combination.

GENERAL CHARACTERISTICS OF
IMMUNOSUPPRESSION IN THE RODENT DISEASE

Rodents infected with trypanosomes or plasmodia have reduced responses to other infections (12) and to sheep red blood cells.(13, 14) Gross changes in the immune system parallel those in the human: splenomegaly, increases in serum immunoglobulin, proliferation of the reticuloendothelial system. Recent experiments have demonstrated a lack of responsiveness to several mitogens and antigens and loss of various other immune functions including cell-mediated immunity, although contact hypersensitivity is only gradually affected. These changes will be discussed in more detail below.

TABLE I

COMMON LABORATORY MALARIA AND TRYPANOSOME INFECTIONS

Causative organism	Host	Duration and severity of infection	Outcome
T.b. brucei	Mouse	Acute. A few days or weeks depending on host and parasite strain.	Fatal
	Rat	Fluctuating low parasitaemia. A few weeks or months.	Fatal
	Rabbit		
	Bovine	Months or years; severe to assymptomatic depending on breed.	Species dependent
T.b. gambiense	Man	Chronic; sometimes many years.	Not always fatal
T. congolense (blood only)	Mouse	A few weeks.	Fatal
T. musculi	Mouse	Constant parasitaemia for 2 weeks.	Non-sterile immunity
P.b. berghei	Mouse		
P.b. yoellii	Young mouse	Acute. Rising parasitaemia. About 2 weeks.	Fatal
P.b. berghei	Young rat		
P.b. yoellii	Adult mouse	Recovery after 3 weeks of parasitaemia.	Non-sterile immunity
P.b. berghei	Adult rat		
P.b. yoellii followed by P.b. berghei	Adult mouse	Chronic infection, fluctuating parasitaemia.	Not fatal
P. malariae	Man	Many cycles of erythro- and exo-erythrocytic stages.	Can be fatal

Parasite location: T. congolense: blood only; other trypanosomes blood and tissue fluids
Plasmodia: intraerythrocytic

COMPONENTS OF THE IMMUNE RESPONSE
WHICH ARE TARGETS FOR IMMUNOSUPPRESSION

Many cell types are involved in the mature immune response.
These different populations affect the various types of response,
and are also functionally interrelated so that malfunction of one
type can influence the response of others. Immunosuppression might
therefore affect a cell population directly or indirectly. I shall
discuss the cells responsible for immunity under the following
headings:

1. "B" lymphocytes (humoral antibody response).

2. "T" lymphocytes ("cell-mediated" immunity; regulation of
    B-cell function).

3. Reticulo-endothelial system (antigen presentation; B-T
    cooperation; release of various chemical mediators).

4. Other factors.

"B" lymphocytes

"B" lymphocytes (B cells) are the cells which are stimulated
by antigen to become antibody-secreting plasma cells, so the
impairment of humoral responsiveness by trypanosomiasis and malaria
could be through primary B-cell insufficiency. A number of
observations suggest abnormal B-cell function. Very high
immunoglobulin (Ig) concentrations are found in the sera of
trypanosome-infected humans,[15] primates,[16] cattle,[17] rabbits
[18] and mice.[19] Although some of these immunoglobulins are
directed against the trypanosome, most are not and will react with
a wide variety of antigens including autoantigens.[16,18] Serum
immunoglobulins are also elevated in malaria.[20,21] IgM can be
ten times normal concentrations; the IgG increase seems to be more
dependent on host-parasite combination; for example, it is raised
five-fold in mice infected with Trypanosoma brucei,[19] but very
little, if at all, in monkeys with Trypanosoma gambiense.[16]

In addition, histological evidence, including use of specific
surface and internal markers, suggests B-cell proliferation with
increases in numbers and maturation into plasma cells in the spleen
and lymph nodes.[8,22,23,24,25,26] (Complement-receptor bearing
lymphocytes in Plasmodium berghei-infected rats are an exception.)
[27] These observations led Greenwood [21] to suggest that B cells
are a target for a mitogen induced or produced by the parasites.

B-cell mitogens, such as bacterial lipopolysaccharide, induce
their target cells to divide following a "programme" of division and

differentiation into antibody secreting cells.(28) The
proliferative capacity of a B-cell clone is limited (28) so that a
clone which has been induced to differentiate is made refractory to
further stimuli. Thus the cells of mice treated with B-cell
mitogens differentiate and produce antibodies of a broad range of
specificity, while becoming unresponsive to newly introduced
antigens.(29,30,31) However, if the B-cell mitogen is given at the
same time as antigen, the immune response is enhanced.(30)

Many observations of B-cell malfunction during trypanosomiasis
and malaria mimic these effects of B-cell mitogens. Apart from the
increase in Ig levels and B and plasma cells already mentioned,
proliferation of spleen cells has been shown in both diseases.(32,
33) In the case of trypanosome infection, B-cell division (23,24)
and immunoglobulin secretion (33) have been directly demonstrated.
Hudson et al. (34) found an increase in the number of "background"
IgM (and sometimes IgG) plaque-forming cells to a variety of
antigens in T. brucei-infected mice and similar observations were
made for malaria.(35) Both parasites show adjuvant properties if
antigen is given one or two days before or after infection is
initiated.(34,20,36) If antigen is given later in infection, both
in vitro (36,37) and in vivo (30,34,38) immune responses are
severely impaired. Moreover, B cells from mice with trypanosomiasis
are unable to respond to the B-cell mitogen bacterial
lipopolysaccharide, even if suppressive cell populations are absent,
(19,33) and other B-cell functions such as T-dependent and
-independent memory are also lost.(19) Lack of mitogen
responsiveness was also demonstrated in P. berghei-infected rats.
(39)

Although in vivo evidence points to a "trypanosome mitogen",
in vitro results are controversial. Both we and Mansfield et al.
(40) have been unable to repeat the observations of Esuruoso (41) of
mitogenicity of trypanosome extracts for mouse spleen cultures.
However, Mansfield et al. (40) could show mitogenicity for rabbit
peripheral blood lymphocytes. Assoku and Tizard (42) found that
autolysates of T. congolense are mitogenic for mouse spleen; their
success could be attributable to the absence of heterologous serum
in the culture medium, as many batches of foetal calf serum are
stimulatory so might mask the mitogenic properties of trypanosomes.
Their trypanosome preparations were probably highly degraded, in
contrast with those tested by other workers.

In the case of malaria, Greenwood and Vick (43) showed that
homogenates of parasitized erythrocytes are mitogenic for mouse
spleen and human peripheral blood lymphocytes; cells from immune

individuals show greater stimulation. Freeman and Parish (44) also found that similar homogenates and their high speed supernatants increase "background" plaque-forming cells to SRBC in spleens of BALB/c mice; theirs are the only experiments where action specifically on B cells has been measured as all others assayed thymidine incorporation.

In summary, it seems highly probable that these parasites are mitogenic for B cells and that this could contribute to immunosuppression. However, other spleen cells are also stimulated to divide during infection (24) and there is no evidence that the action on B cells is direct.

"T" lymphocytes

"T" lymphocytes (or "T cells") are a heterogeneous population of cells which have undergone a period of maturation in the thymus and which bear the Thy 1 (θ) surface marker. The different T-cell subsets have widely differing functions, so it is not surprising that they are affected in a variety of ways by trypanosomes.

The initial effect of infection on the T cells of mouse spleen is a stimulation of division which results in a transient increase in T cell numbers.(24) However, the number of T cells subsequently decreases,(24) the T-dependent areas being replaced by plasma cells and macrophages.(22,45) Histological changes in the malarious spleen are similar, but the effect on T cell numbers seems to depend on the host-parasite combination used. Young rats with fatal P. berghei infection show thymic involution and decreased T cell numbers in the spleen, but in older rats which are able to recover the decrease is transient and slight.(27) Jayawardena et al. (46) found that the proportion of splenic dividing cells which bore a chromosome T-cell marker increased markedly during Plasmodium yoellii, but only slightly during fatal P. berghei infection in mice Unfortunately, these observations are uninterpretable without knowledge of the proportion or number of spleen cells which were dividing. Krettli and Nussenzweig (47) showed a decrease in the number of T cells in thymus and both T and B cells in the lymph nodes of P. berghei-infected mice. In general, there seems to be a positive correlation between the decrease in T cell numbers in T. brucei and P. berghei infections and the fatality of the disease.

Several T-cell functions are severely depressed during trypanosome infection. The responses to the T-cell mitogens PHA (phytohaemagglutinin) and Con A (concanavalin A) are diminished (33, 48) and there is loss of memory for the T-cell ability to cooperate with B cells in the humoral response.(19) In addition, T cells become unable to react to heterologous H-2 antigens in the mixed lymphocyte reaction.(19,37) Although some of the depression of these responses was due to other suppressive cells,(37,48) the intrinsic capability of the T cells being investigated was definitely diminished. For instance T-helper memory remains deficient even if suppressive T cells and macrophages are removed by two passages of the responding population through nylon wool. (19) In contrast, the depression of the Con A and PHA responses in murine P. berghei infection can be abrogated by removal of suppressive macrophages.(49)

Despite the loss of some T-cell functions during trypanosomiasis, others are relatively unaffected or even enhanced. Contact hypersensitivity is only depressed in the terminal stages of infection in mice (22) and probably also rabbits.(50) The picture in malaria is similar, as P. yoellii infection of mice does not affect skin graft rejection or contact hypersensitivity (14) and peripheral blood lymphocytes of malarious children respond normally to PHA.(6) These experiments were complicated by an increase in "background" thymidine uptake by the cells from infected patients.

I have already described how the stimulation of B cells can lead to a state of unresponsiveness, and similar considerations may apply to the subsets of T cells which become deficient in trypanosomiasis. However, just as activation of B cells leads to immunoglobulin secretion, the action of mitogens on one subset of T cells can result in the expression of function - that of active suppression of the responses of other T and B cells.(51)

In accordance with this, suppressor T cells have been demonstrated in trypanosomiasis. Jayawardena and Waksman (48) and Eardley and Jayawardena (37) have demonstrated a population of cells in the spleen of T. brucei-infected mice which suppresses the response of normal mouse spleen cells to mitogens and sheep red blood cells. Some of these were "adherent" cells, the rest being T cells. Their experiments were performed at in vitro cell densities of 7.5 x 10^6 per ml or more. In contrast, we have not found any suppression of the mitogen responsiveness of normal cells by cells from infected animals, at 2 x 10^6 cells per ml or less. We could, however, restore responsiveness to LPS by removing the T cells from spleen cell populations taken during the early stages of trypanosome infection.(33,19) It is important to remember that lymphocytes from trypanosome-infected animals show very poor survival in culture, so some of the suppressive activity at high cell densities may be attributable to toxicity from large numbers of dying cells.

Although suppressive T cells contribute to immunosuppression, they are unlikely to be wholly responsible as immunosuppression also occurs in infected T-deprived (19) and nude mice (Clayton, unpublished results).

The reticulo-endothelial system

The reticulo-endothelial system shows intense proliferation during trypanosomiasis and malaria.(5,8,45,10,52) This is seen histologically and in an increased rate of clearance of intravenously injected particles. However, the proportion of particles taken up by the spleen is not increased despite splenomegaly (45,52) and the clearance rate depends on both administration route (52) and the host animal.(53) Both infections cause anaemia (see below), with extensive erythrophagocytosis which could be a consequence of red cell damage or opsonization.(10,45,54 55)

Macrophages have many roles in the immune response. They enhance B - T cooperation (56) by "presentation" of the antigen,(57) and release lymphokines, complement components and prostaglandins, besides being capable of direct killing of other cells.(58) Disruption of any of these functions could impair immunity.

Macrophages have been clearly implicated in malarial immunosuppression. Various lines of evidence initially suggested a failure at the level of T - B cooperation: children with acute malaria had decreased humoral responses to the T-dependent "O" antigen but not the T-independent "H" antigen of S. typhii;(6) mice at the peak of P. yoellii parasitaemia respond to $\emptyset$ x 174 and DNP-Ficoll (both "T-independent") but not sheep red cells (mostly "T-dependent").(35,36) Loose et al. (59) suggested that the impaired capacity of macrophages for endotoxin detoxification during malaria infection (60) might extend to antigen processing. They showed that peritoneal cells from infected animals were defective in antigen presentation,(59) although suppressive cells were not ruled out. Later work has shown that the in vitro responsiveness to antigen of spleen cells from P. yoelli-infected mice can be restored by removing the adherent cells and adding back adherent cells from normal mice.(36) As active suppression by the macrophages from infected mice is not demonstrable,(61) a defect in macrophage antigen processing is probably of major importance in mice infected with P. yoellii. However, with P. berghei the situation is different: the spleen cells of infected rats are still unresponsive to mitogens after adherent cell removal (38) and the macrophages from infected mice actively suppress lymphocyte responses to mitogens.(49)

During T. brucei infection in mice, antigen handling by macrophages is not defective (45) but macrophages can be suppressive in vitro (33, 48) and in vivo (Corsini, unpublished results). The alterations in macrophage function could be a consequence of phagocytosis of large numbers of parasites or erythrocytes.(62)

It is clear from the above discussion that no observations made of immunosuppression are interpretable without precise definition of the cell populations involved. Purification procedures must be carefully monitored: a ninety per cent removal of a highly active population may be inadequate; also, cells from infected animals are often deranged, rendering standard separation methods ineffective. For instance, the density of surface immunoglobulin is decreased on B cells from mice with trypanosomiasis.(24) The increased fragility of cells from infected animals, especially in culture, is another hazard.

Other factors

Many factors apart from the cell populations considered above play a role in immunity. Lymphokines are probably involved, for instance in attracting macrophages to the spleen.(63) Disruption of spleen and lymph node architecture and of lymphocyte recirculation could impair cell-cell cooperation,(22,25) and other pathological effects, such as oedema, inflammation and complement depletion,(64) could result in immunosuppression, as could any stress caused by the disease.(65)

Both malaria and trypanosomiasis result in severe anaemia. The possible causes of the anaemia in trypanosomiasis seem to vary in importance with the host animal, and include damage by a toxic trypanosome component,(66) which increases erythrocyte fragility, (67) and autoimmune damage caused by autoantibodies induced by the parasite,(68) both of which result in erythrophagocytosis.(67) The animal responds to the anaemia by reticulocytosis and normoblastic hyperplasia of the erythropoietic organs.(67) As well as disrupting splenic architecture (25) this erythropoiesis can affect the supply of lymphoid cells from bone marrow. Various manipulations of the balance between erythrocytes, granulocytes and leucocytes suggest that these three cell types have a common stem cell precursor. Anaemic animals are immunosuppressed and more susceptible to infection, whereas animals in rebound from hypoxia are hyper-responsive.(69,70)

AREAS FOR FURTHER INVESTIGATION

Possible factors operating during immunosuppression are summarized in Table II. There are some for which there is a considerable body of evidence, but others are predominantly speculative. There are several areas which might be profitably explored:

1)   The macrophage

Although phagocytosis and antigen presentation by macrophages has been investigated, relatively little attention has been paid to other macrophage functions involved in the immune response, such as release of mediators. During malaria and trypanosomiasis, the macrophages show "activated" appearance with abundant cytoplasm, suggesting that, for instance, secretion might be enhanced.

2)   What are null cells?

We have recently shown that most of the cells stimulated by trypanosomes to divide in the spleen bear neither B nor T-cell markers.(23,24) These cells could be haemopoietic, reticulo-endothelial, early undifferentiated lymphoid, or some other as yet undefined cell type. They eventually form the majority of spleen cells in infected rodents, so could be extremely important in immunosuppression. One implication is that any remaining responsiv B and T cells will be diluted. Analogous cells are seen in the spleen after destruction of lymphocytes with cyclophosphamide,(71) so the "null" cells in trypanosomiasis could be a response to similar destruction.

3)   Role of autoantibodies

Some of the autoantibodies induced by trypanosomes might immunosuppress by being toxic to lymphocytes.

4)   Relationship between course of infection and immunosuppression

As noted above, parasites which are lethal for their host cause more rapid, severe immunosuppression than those which the hos

TABLE II

POSSIBLE FACTORS IN IMMUNOSUPPRESSION

A. __Pathological symptoms__

      1)   Splenomegaly, lymph-node enlargement.

      2)   Disruption of lymphoid architecture.

      3)   High serum immunoglobulin.

B. __B cells__

      Proliferation, spontaneous Ig secretion leading to
unresponsiveness.

C. __T cells__

      Proliferation?

      Unresponsiveness (some subsets).

      Activation of suppressive T cells.

D. __Macrophages__

      Activated appearance, erythrophagocytosis.

      Impaired antigen handling/presentation.

      Suppressive activity.

      Release of chemical mediators.

E. __Other factors__

      Complement depletion.

      Anaemia; competition for stem cells in bone marrow.

      Parasite factors: toxic, mitogenic.

      Autoantibodies.

      Null cells.

can control. (Compare P. berghei and T. brucei with P. yoellii and
T. musculi in mice.) These situations are difficult to analyse as
the more lethal infections result in very high parasitaemias, which
by themselves could cause the difference in degree of
immunosuppression. Variations in susceptibility between inbred
mouse strains could be useful in investigating this phenomenon,(23,
72) but ideally systems need to be developed where the quantity of
parasite material is similar in the animals being compared.

An example of the insight which can be gained by comparing
infections is that T. brucei raises IgM and IgG levels and
depresses both types of response in mice, whereas T. musculi raises
only IgG and only significantly depresses IgG responses (Clayton,
unpublished results). This suggests a direct link between non-
specific activation of a population and suppression of its
response. Direct comparison of various infections and antigenic
competition situations could also reveal common underlying
mechanisms. Large quantities of antigen can by themselves compete
for macrophages,(62) suppress responses to mitogen (73) and lead to
production of non-antibody immunoglobulins;(74) and enormous
antigen loads are certainly present during infections.

THE IMMUNOSUPPRESSIVE AGENT

This is usually identified either as a toxic parasite product
(75) such as tryptophol (76) or free fatty acids,(77) or the
postulated mitogen. If there is a mitogen, it is the most
polyspecific one known, but so far there is no evidence whether any
parasite product or component acts directly on many cell types, or
on one type which then influences others. If the mitogenic and
toxic parasite components could be reliably assayed, purified and
characterized, then their relationship with immunosuppression might
be clarified.

CONCLUDING REMARKS

Much is now known about the phenomenology of immunosuppression
in malaria and trypanosomiasis, but we are still in the dark about
the underlying mechanism. If one specific immunosuppressive
molecule is identified then "immunization against immunosuppression"
might be feasible. However, I suspect that antigenic load, toxic or
mitogenic metabolic products or components of the parasite,
autoimmunity, immune complexes and anaemia probably all play a part.

REFERENCES

1. SHUMM, D.E. and BILLMIRE, D.F. Progressive loss of in vitro immune response with tumour growth. Cellular Immunology, 24: 348 (1976).

2. BRITO, I.V. ET AL. Immunological response to tetanus toxoid during a schistosomal infection in mice. Journal of Tropical Medicine and Hygiene, 79: 161 (1976).

3. OTTERNESS, I.G. and CHANG, Y-H. Evidence for suppressor cell activity associated with depression of contact sensitivity in Pseudomonas aeruginosa infected mice. Clinical and Experimental Immunology, 26: 355 (1976).

4. SELGRADE, M.K. ET AL. Effect of Murine Cytomegalovirus on the in vitro responses of T and B cells to mitogens. Journal of Immunology, 116: 1489 (1976).

5. MOTT, F.W. Histological observations on sleeping sickness and other trypanosome infections. In: Report of the Sleeping Sickness Commission No. VII. Royal Society of London (1906).

6. GREENWOOD, B.M. ET AL. Immunosuppression in children with malaria. Lancet, i: 169 (1972).

7. GREENWOOD, B.M. Immunosuppression in trypanosomiasis. In: Parasites in the Immunised Host: Mechanisms of survival. Ciba Foundation Symposium 25. Elsevier-North Holland, Associated Scientific Publishers, Amsterdam, p.137 (1974)

8. GREIG, E.D.W. and GRAY, A.C.H. Continuation report on sleeping sickness in Uganda. In: Report of the Sleeping Sickness Commission No. VI. Royal Society of London (1905).

9. APTED, F.I.C. Clinical manifestations and diagnosis of sleeping sickness. In: The African Trypanosomiases (Mulligan, H.W. ed.) Allen and Unwin, London, Chapter 35 (1970).

10. EDINGTON, G.M. and GILLES, H.M. Protozoal diseases. In: Pathology in the Tropics. Second edition, Arnold, London, p.10 (1976).

11. SCOTT, J.M. ET AL. Immunosuppression in bovine trypanosomiasis: field studies using foot-and-mouth disease vaccine and clostridial vaccine. Tropical Animal Health and Production, 9: 159 (1977).

12. PHILLIPS, R.S. ET AL. The effect of Plasmodium berghei and Trypanosoma brucei infections on the immune expulsion of the nematode Trichunis muris from mice. International Journal of Parasitology, 4: 409 (1974).

13. GOODWIN, L.G. ET AL. Immunosuppression during trypanosomiasis. British Journal of Experimental Pathology, 53: 40 (1972).

14. GREENWOOD, B.M. ET AL. Immunosuppression in murine malaria. I.General characteristics. Clinical and Experimental Immunology, 8: 467 (1971).

15. MASSEYEFF, R. and LARRY, J. Serum immunoglobulin levels in African Trypanosomiasis. Clinica Chimica Acta, 14: 285 (1966).

16. HOUBA, V. ET AL. Heterophile antibodies, M-anti-globulins and immunoglobulins in experimental trypanosomiasis. Clinical and Experimental Immunology, 4: 113 (1969).

17. KOBAYASHI, A. and TIZARD, R.R. The response to Trypanosoma congolense infection in calves. Tropenmedizin und Parasitologie, 27: 411 (1976).

18. MANSFIELD, J.M. and KREIER, J.P. Autoimmunity in experimental T. congolense infections of rabbits. Infection and Immunity, 5: 648 (1972).

19. ASKONAS, B.A. ET AL. Functional depletion of T- and B- memory cells and other lymphoid cell subpopulations during trypanosomiasis. Immunology (in Press).

20. POELS, L.G. and VAN NIEKERK, C. Plasmodium berghei: Immunosuppression and hyperimmunoglobulinemia. Experimental Parasitology, 42: 235 (1977).

21. GREENWOOD, B.M. Possible role of a B cell mitogen in hypergammaglobulinaemia in malaria and trypanosomiasis. Lancet, 1: 435 (1974).

22. MURRAY, P.K. ET AL. The nature of immunosuppression in Trypanosoma brucei infections in mice. II. The role of the T and B lymphocytes. Immunology, 27: 825 (1974).

23. MORRISON, W.I. ET AL. Susceptibility of inbred strains of mice to Trypanosoma congolense: correlation with spleen lymphocyte populations. Clinical and Experimental Immunology, 32: 25 (1978).

24. MAYOR WITHEY, K.S. ET AL. Trypanosomiasis leads to extensive proliferation of B- T- and null cells in spleen and bonemarrow. Clinical and Experimental Immunology (submitted for publication).

25. MORAN, C.J. ET AL. The immunological significance of histological changes in the spleen and liver in mouse malaria. Clinical and Experimental Immunology, 13: 467 (1973).

26. WEDDERBURN, N. ET AL. Chronic malarial infection in Balb/c mice. Effect on the immune response to sheep erythrocytes and histological changes in the liver and spleen. Transactions of the Royal Society of Tropical Medicine and Hygiene, 69: 468 (1975).

27. GRAVELY, B.M. ET AL. T and B cell population changes in young and adult rats infected with Plasmodium berghei. Infection and Immunity, 14: 178 (1976).

28. ASKONAS, B.A. and NORTH, J.R. The lifestyle of B cells. Cold Spring Harbor Symposium on Quantitative Biology, p.749 (1977).

29. BRALEY-MULLEN, H. ET AL. Suppression of IgM responses to Type III Pneumococcal polysaccharide by lipopolysaccharide (LPS). Cellular Immunology, 30: 300 (1977).

30. NAKASHIMA, I. ET AL. Alterations in the antibody response to bovine serum albumin by capsular polysaccharide of Klebsiella pneumoniae. Journal of Immunology, 107: 1112 (1971).

31. NAKASHIMA, I. and KATO, N. Non specific stimulation of immunoglobulin synthesis in mice by capsular polysaccharide of Klebsiella pneumoniae. Immunology, 27: 179 (1974).

32. GOLENSER, J. ET AL. Dynamics of thymidine incorporation by spleen cells from rats infected with Plasmodium berghei. Clinical and Experimental Immunology, 22: 364 (1975).

33. CORSINI, A.C. ET AL. Suppressor cells and loss of B cell potential in mice infected with Trypanosoma brucei. Clinical and Experimental Immunology, 29: 122 (1977).

34. HUDSON, K.M. ET AL. Immunodepression, high IgM levels and evasion of the immune response in murine trypanosomiasis. Nature, 264: 256 (1976).

35. BARKER, L.R. Experimental malaria: effects upon the immune response to different antigens. Journal of Infectious Diseases, 123: 99 (1971).

36. BROWN, I.N. ET AL. Antibody response in vitro of spleen cells from Plasmodium yoellii-infected mice. Infection and Immunity, 16: 456 (1977).

37. EARDLEY, D.D. and JAYAWARDENA, A.N. Suppressor cells in mice infected with Trypanosoma brucei. Journal of Immunology, 119: 1029 (1977).

38. WEDDERBURN, N. Immunodepression produced by malarial infection in mice. In: Parasites in the Immunized Host. Ciba Foundation Symposium 25. Elsevier-North Holland, Associated Scientific Publishers, Amsterdam, p.123 (1974)

39. SPIRA, D.T. ET AL. The reactivity of spleen cells from malarious rats to non-specific mitogens. Clinical and Experimental Immunology, 24: 139 (1976).

40. MANSFIELD, J.M. ET AL. Lymphocyte function in experimental African trypanosomiasis: mitogenic effects of trypanosome extracts in vitro. Infection and Immunity, 14: 976 (1976).

41. ESURUOSO, G.O. The demonstration in vitro of the mitogenic effects of trypanosomal antigen on the spleen cells of normal athymic and cyclophosphamide treated mice. Clinical and Experimental Immunology, 23: 314 (1976).

42. ASSOKU, R.K.G. and TIZARD, I.R. Mitogenicity of autolysates of Trypanosoma congolense. Experientia, 34: 127 (1978).

43. GREENWOOD, B.M. and VICK, R.M. Evidence for a malaria mitogen in human malaria. Nature, 257: 592 (1975).

44. FREEMAN, R.R. and PARISH, C.R. Polyclonal B cell activation during rodent malarial infections. Clinical and Experimental Immunology, 32: 41 (1978).

45. MURRAY, P.K. ET AL. The nature of immunosuppression in Trypanosoma brucei infections in mice. I. The role of the macrophage. Immunology, 27: 815 (1974).

46. JAYAWARDENA, A.N. ET AL. T cell activation in murine malarias. Nature, Lond. 258: 149 (1975).

47. KRETTLI. A.U. and NUSSENZWEIG, R. Depletion of T and B lymphocytes during malarial infections. Cellular Immunology, 13: 440 (1974).

48. JAYAWARDENA, A.N. and WAKSMAN, B.H. Suppressor cells in experimental trypansomiasis. Nature, 265: 539 (1977).

49. JAYAWARDENA, A.N. T cell dependent effector and suppressor responses in experimental malaria. 5th International Congress of Protozoology, Abstract 72 (1977).

50. MANSFIELD, J.M. and WALLACE, J.H. Suppression of cell mediated immunity in experimental African trypanosomiasis. Infection and Immunity, 10: 335 (1974).

51. DUTTON, R.W. Suppressor T cells. Transplantation Reviews, 26: 39 (1975).

52. GREENWOOD, B.M. ET AL. Immunosuppression in murine malaria. II. The effect on reticulo-endothelial and germinal centre function. Clinical and Experimental Immunology, 9: 345 (1971).

53. CANTRELL, W. and ELKO, E.E. Plasmodium berghei: phagocytic activity in two strains of rats. Experimental Parasitology, 40: 281 (1976).

54. JENNINGS, R.W. ET AL. Anaemia in trypanosomiasis: studies in rats and mice infected with T. brucei. Research in Veterinary Science, 16: 70 (1974).

55. MACKENZIE, P.K.I. and CRUIKSHANK, J.G. Phagocytosis of erythrocytes and leukocytes in sheep infected with Trypanosoma congolense. Research in Veterinary Science, 15, 256 (1973).

56. ERB, P. and FELDMANN, M. The role of macrophages in the generation of T-helper cells. III. Influence of macrophage-derived factors in helper cell induction. European Journal of Immunology, 5: 759 (1975).

57. ASKONAS, B.A. and ROELANTS, G.E. Macrophages bearing hapten-carrier molecules as foci inducers for T and B lymphocyte interaction. European Journal of Immunology, 4: 1 (1974).

58. NELSON, D.S. Immunobiology of the macrophage. Academic Press, New York, San Francisco, London (1976).

59. LOOSE, L.D. ET AL. Malarial immunosuppression - a macrophage mediated defect. Proceedings of the Helminthological Society of Washington, 39 (Special issue): 484 (1972).

60. LOOSE, L.D. ET AL. Impaired endotoxin detoxification as a factor in enhanced endotoxin sensitivity of malaria-infected mice. Proceedings of the Society for Experimental Biology and Medicine, 137: 794 (1971).

61. WARREN, H.S. and WEIDENZ, W.P. Malarial immunosuppression in vitro: adherent spleen cells are functionally defective as accessory cells in the response to horse erythrocytes. European Journal of Immunology, 6: 816 (1976).

62. SABET, T. ET AL. Effects of reticulo-endothelial system "blockage" in antibody formation. I. Suppressed cellular and humoral haemolysin responses in mice injected with carbon particles. Immunology, 16: 433 (1969).

63. WYLER, D.J. and GALLIN, J.I. Spleen-derived mononuclear cell chematactic factor in malaria infection: a possible mechanism for splenic macrophage accumulation. Journal of Immunology, 118: 478 (1977).

64. GREENWOOD, B.M. and WHITTLE, H.C. Complement activation in patients with Gambian sleeping sickness. Clinical and Experimental Immunology, 24: 133 (1976).

65. MORGAN, A.A. and COLLECTOR, M.I. Stress induced modulation of the immune response. Science, 196: 307 (1977).

66. CHI, N.H. ET AL. Haemolytic effects of Trypanosoma brucei. In: Biochemistry of Parasites and Host Parasite Relationships. Elsevier-North Holland Biomedical Press, p.409 (1976)

67. JENNINGS, F.W. ET AL. Anaemia in trypanosomiasis: studies in rats and mice infected with T. brucei. Research in Veterinary Science, 16: 70 (1974).

68. KOBAYASHI, A. ET AL. Studies on the anaemia in experimental African trypanosomiasis. II. The pathogenesis of the anaemia in calves infected with Trypanosoma congolense. American Journal of Tropical Medicine and Hygiene, 25: 401 (1976).

69. FRANKENBERG, S. ET AL. Plasmodium berghei: relationship between mitosis and erythropoiesis in spleen cells of infected rats. Experimental Parasitology, 38: 105 (1975).

70. GREENBLATT, C.L. and YOFFEY, J.M. Trypanosoma lewisi: Immunohaematopoietic interrelationships of the infection in normal, hypoxic and rebound animals. Experimental Parasitology, 38: 105 (1975).

71. KOLB, J-P.B. ET AL. Splenic modifications induced by cyclophosphamide in C3H/He, nude, and "B"- mice. Journal of Immunology, 118: 1595 (1977).

72. CLAYTON, C. T.b. brucei: effect of host strain and parasite antigenic type on infection in mice. Experimental Parasitology, 44: 202 (1978).

73. GERSHON, R.K. ET AL. Suppressive effects of in vivo immunisation on PHA response in vitro. Journal of Immunology, 112: 215 (1974).

74. ANTOINE, J.C. ET AL. Development of immunoglobulin and antibody-synthesising cells after immunisation with different doses of antigen. Immunology, 31: 921 (1976).

75. TIZARD, I.R. and HOLMES, W.L. The generation of toxic activity from Trypanosoma congolense. Experientia, 32: 1533 (1976).

76. SEED, J.R. and HALL, J.E. The possible role of the trypanosome metabolite indole - 3 - ethanol in the neuropathology of trypanosomiasis. 5th International Congress of Protozoology, Abstract 115 (1977).

77. ASSOKU, R.K.G. ET AL. Free fatty acids, complement activation, and polyclonal B-cell stimulation as factors in the immunopathogenesis of African trypanosomiasis. Lancet, 11: 956, (1977).

ACKNOWLEDGEMENTS

I wish to thank my supervisors Drs B.A. Askonas and B.M. Ogilvie for generous help and advice, and the Medical Research Council (Great Britain) for their support.

DISCUSSION

MORRIS: What happens to antigen load in trypanosomiasis?

CLAYTON: Well, one assumes that the organisms are taken up by macrophages, either after opsonization or perhaps after lysis. In the descending phase of a peak of parasitaemia, agglutinated organisms and ones with low motility and rather swollen appearance are seen, and immunity is thought to be B-cell mediated. I personally can't remember seeing any paper in which pagocytosis of trypanosomes has been proved by serial E.M. sections, although Takayanagi and Enriquez say it occurs and adherence is well documented. Fluorescent antibody to the trypanosome surface coat reveals patches of trypanosome material apparently inside infected mouse peritoneal macrophages but again this could be stuck to the surface and Jack Doyle says that they phagocytose the surface coat and leave the trypanosome outside. However, phagocytosis and subsequent digestion seems the most probable fate of the parasite material.

HUMPHREY: When you speak of immunosuppression due to macrophages later in infection are you referring to adherent cells from the spleen? If so, does this involve only those macrophages (MØ) which adhere to glass/plastic? Or do you know whether the quite large proportion of non-adherent MØ are different in this respect?

CLAYTON: Our immunosuppressive macrophages were adherent cells from the peritoneum; those of Jayawardena's group were from the spleen, and I am deducing that they were macrophages from the fact that they adhered to glass wool or nylon wool. I don't know whether non-adherent macrophages are different in this respect - it's an intriguing possibility that the suppressive and adherent properties might be correlated.

NOGUEIRA: What proportion of the peritoneal cells in your infected mice were macrophages?

CLAYTON: Roughly half (by neutral red uptake), so that the increase in peritoneal macrophages was accompanied by an equal infiltration with (morphologically) small lymphocytes. Perhaps this infiltration is an example of that seen histologically in many other sites - such as the brain in sleeping sickness? Tissue infiltration could also be responsible for the dramatic drop in leucocytes in the peripheral blood late in infection.

MITCHELL: Are there any of your observations which do not fit with the following scheme? Massive extramedullary haemopoiesis, largely stimulated by factors derived from, or dependent upon, hyperactive macrophages in large numbers, results in depletion of stem cells from the bone marrow and a dilution of antigen-reactive cells by normoblasts in the spleen. Rob Freeman in Canberra has shown that, in murine malaria, there is an enormous increase in normoblasts and that the resulting dilution of immunologically active cells is a sufficient reason for the reduced immunological activity of a fixed number of mononuclear cells in vitro. One may also have to be careful about denoting a surface Ig+ cell as a B cell in splenomegalic situations. Other "lymphocyte" markers should also be used with caution since we have recently shown that mouse eosinophils are strongly Ly2+.

CLAYTON: The haemopoiesis and dilution by normoblasts must certainly play a role. I think it is stimulated as a response to anaemia, perhaps caused by parasite-derived haemolytic factors, as well as by factors from hyperactive macrophages. We tend towards the view that dilution of immunologically-reactive cells per se is insufficient to account for immunosuppression. For instance, at two weeks of infection roughly a third of the cells in our spleen cell suspensions are B cells, the rest being "null". However, the response of these B cells to secondary stimulation with DNP-Ficoll in adoptive transfer is totally abolished, as are many other responses. I agree wholeheartedly about the need for caution when looking at cells from these enlarged spleens. However, I think that to an extent it is a question of experience. I have looked at an enormous number of preparations stained by immunofluorescence and I am sure that I can tell the difference between a B cell which has a characteristic pattern of strong surface staining and, say, a macrophage which shows dispersed dots of fluorescence as a consequence of adherence via Fc receptors.

COHEN: Is there in vivo evidence for B cell mitogens acting early in vivo e.g. increased IgG levels?

CLAYTON: In our mice, IgG levels rise at about 10 days, and IgM
very slightly later. The precise onset of the "mitogenic" effect
is difficult to measure because the number of parasites present is
changing. For the first few days after infection (we usually
infect with 200 parasites) very few parasites are present so one
would not expect to see any effect. However, spontaneous 3-H
thymidine incorporation is already increased by day 5 of infection
so one assumes that by day 3 or 4 sufficient parasites are present
to be stimulating (perhaps helped by the fact that T. brucei S42
tends to localize preferentially in the spleen). If the initial
injected dose is increased to, say, a million, then the
stimulation of thymidine incorporation is measurable by day 3.

GREENWOOD: Have you tested immunosuppressed mice for enhanced K
cell activity? Could some of the increased number of null cells
you have observed be K cells?

CLAYTON: There is someone working on this at the moment but so
far we have no definite answer.

POLTERA: Animal and human trypanosomiasis may run a chronic form
of disease. Your experiments are covering the first 3-4 weeks of
infection. Can you comment on a chronic experimental model?

CLAYTON: The most chronic mouse models are T. congolense, and
T. brucei where immunosuppression still occurs (Roelants, Pearson,
Hudson and Terry, personal communication), as it does in the rabbit
model which is also fairly chronic. One does have to be very
careful in extrapolating from the relatively acute models to the
field situation. In general, the more chronic experimental systems
show a more gradual onset of immunosuppression than the acute ones,
so immunosuppression probably takes even longer to develop in man
and cattle with a disease of months' or years' duration. Possibly
there is a positive correlation between parasite load and
immunosuppression - an attractive proposition if one favours the
mitogen hypothesis. I think it would be true to say that, broadly
speaking, more chronic diseases manifest lower parasitaemias,
although when parasites are in tissue fluids as well as blood,
precise quantitation is impossible. The overall symptoms in
rodents and large mammals (splenomegaly, high IgM, etc.) are in fact
similar, suggesting that our rodent experiments will be of more
general applicability.

FAKUNLE: Dr Clayton, you mentioned the generation of T suppressor
cells. Has there been quantitation of helper and suppressor cell
numbers?

CLAYTON: We know that <u>memory</u> for T help is abolished by two weeks of infection.(<u>19</u>) As far as I know, suppressor cells have not yet been quantitated. It would have to be done by some sort of limiting dilution analysis. This would probably have to be done at high cell density as, so far as I know, active suppression of normal spleen cell responses by the suppressor cells only occurs at overall cell densities greater than 5×10^6 per ml. I do worry a little whether the effects at these densities are due to death of the very fragile cells from infected animals.

MILLER: What happens to the lymph node population in trypanosome-infected mice? Is delayed-type hypersensitivity suppressed?

CLAYTON: In rodents, delayed-type hypersensitivity (in the form of skin testing with a variety of agents) is only suppressed fairly late on, and we think may even be elevated in the early stages.(<u>19</u>) As far as I know, that is the only test of lymph node function which has been done. However, we do know that, like the spleen, lymph nodes are enlarged, and have increased spontaneous cell division (measured by ^{125}IUdk incorporation).(<u>22</u>) Children with trypanosomiasis have reduced DTH.(<u>7</u>) Lymph nodes are one of the neglected areas, and certainly should be investigated.

HUMPHREY: How are you measuring DTH? By local skin reaction, where the visible changes are mediated by recruited blood monocytes, or by MIF production? Could it be that monocytes are able to produce skin swelling even in response to diminished amounts of MIF produced locally by T (helper) cells?

CLAYTON: We are measuring DTH by local skin reaction, although it has also been measured by ^{125}IUdk incorporation into the regional nodes. Where swelling is concerned, it is quite possible that a decrease in T cell reactivity could be more than compensated by an increase in both numbers and responsiveness in the monocyte population.

SPRENT: <u>A propos</u> the role of the marrow in the immunosuppression you see, have you tried using ^{90}Sr-treated mice?

CLAYTON: No, we haven't - it would be very interesting to see how much immunosuppression could be achieved by eliminating the bone marrow.

WEISS: Dr Clayton, you describe a decrease in stem cells in bone marrow, an increase in cell proliferation in the spleen and an

increase in "mononuclear cells" in the peritoneal cavity. What is
the appearance of the bone marrow - is it aplastic? Have you
carried out the Till-McCulloch stem cell assay for spleen and for
the cells in the peritoneal cavity?

CLAYTON: The histological studies were done by Dr John White, who
saw severe structural disruption only a few days after infection.
Most workers report normoblastic hyperplasia. However, I am not a
histologist so am not able to comment in any detail. We have not
tried the Till-McCulloch assay for spleen and peritoneal cells,
but I think the results would be very instructive. It would also
be interesting to look at peripheral blood as there is evidence for
release of stem cells into the circulation in response to anaemia.

Role of the Spleen in Chagas' Disease*

Z. Brener, A.U. Krettli, Department of Parasitology,
University of Minas Gerais and Centro de Pesquisas
René Rachou, FIOCRUZ, Brazil, and A. Alcantara F$^{\underline{o}}$,
School of Medicine, Department of Pathology, Catholic
University of Pelotas, R.G. Sul, Brazil.

The role played by the spleen in Trypanosoma cruzi infections
should be considered from two different aspects: a) as an
"effector immunological organ" (1) which participates in the host
immune response; b) as an organ bearing heavy parasitism in
infection by T. cruzi strains which selectively interact with
cells from the mononuclear phagocytic system.

The existence of a humoral response in Chagas' disease has
been extensively demonstrated by the presence of serum antibodies
in both acute and chronic stages of the infection. Increase of
IgM has been reported in the human acute disease. A significant
increase of total serum gammaglobulins as well as IgM and IgG
types reactive to T. cruzi in experimentally infected mice during
the acute phase has been described.(2) Serological tests
(immunofluorescence, haemagglutination, complement fixation) are
consistently positive in the chronic phase.

The participation of antibodies in the host resistance has
been considered as controversial but recent investigations clearly
demonstrated that resistance can be passively transferred to
normal recipient mice by inoculation of specific antiserum.
Krettli and Brener (3) investigated the action of immune-sera
collected from animals in the chronic phase on bloodstream forms

* This work was supported by a grant from the National Research
Council, Brazil

from two T. cruzi strains (Y and CL). The parasites from the Y strain (which shows predominance of slender trypomastigotes) were intensively agglutinated by anti-CL and anti-Y immune-sera and had their infectivity decreased when inoculated into mice; CL parasites, however, were not agglutinated by either sera and the decline in infectivity was not observed. Mice injected with anti-Y or anti-CL immune-sera were clearly protected against a challenge infection by Y strain bloodstream forms. Hanson et al. (2) also induced protection from a lethal infection in mice by transferring immune sera obtained from mice six weeks after inoculation.

Despite the clear evidence of a participation of antibodies in the host resistance, splenectomy apparently does not significantly affect the course of infection in animals inoculated with T. cruzi. Nieschulz and Wawo-Roentos (4) demonstrated that splenectomized dogs present infections similar to those in intact hosts. The same results were observed using mice, dogs and guinea-pigs.(5) The mild course of infection in rats infected with T. cruzi-like strains from the United States of America was not changed by splenectomy.(6) Splenectomy performed during established infection did not significantly increase parasitaemia and mortality.(7) Pizzi and Knierim (8) described a significant decline in the antibody titres in splenectomized mice as compared with intact ones but this phenomenon had no influence on the course of infection.

The effect of splenectomy upon T. cruzi parasitaemia in experimentally infected mice was recently reinvestigated in our laboratory. About 100,000 bloodstream forms from the Y strain were inoculated by intraperitoneal route in groups of splenectomized sham-splenectomized and intact mice. A slight but not significant increase of the number of parasites was observed in the splenectomized animals but the general course of infection was not affected (Fig. 1). Similar results were obtained with splenectomized mice inoculated by subcutaneous route (Fig.2). Preliminary results with the CL strain show that parasitaemia is higher in splenectomized mice than in normal animals but the general course of infection was not changed.

Krettli and Nussenzweig (9) have recently demonstrated that bloodstream forms of T. cruzi Y strain obtained from acutely infected mice are coated with immunoglobulins. Such parasites are lysed in vitro upon addition of normal human serum as a result of activation of the alternative pathway of complement. Further evidence of the presence of specific immunoglobulins on the

FIGURE 1

CURVES OF PARASITAEMIA IN GROUPS OF SPLENECTOMIZED,
SHAM-SPLENECTOMIZED AND NORMAL CONTROL MICE, INOCULATED WITH
100,000 BLOODSTREAM FORMS OF T. CRUZI Y STRAIN BY INTRAPERITONEAL
ROUTE. INOCULATION THREE DAYS AFTER SPLENECTOMY

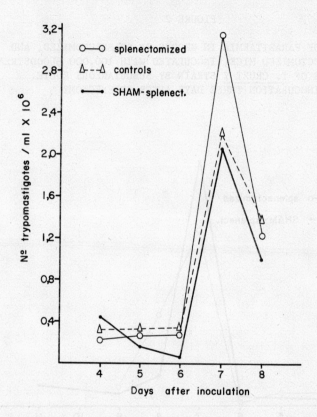

parasites membrane was provided by fluorescein-labelled and
isotope-labelled anti-gammaglobulin. We have now investigated the
presence of immunoglobulins on the surface of parasites collected
from splenectomized and sham-splenectomized mice by complement-
mediated lysis according to the technique described by Krettli
et al. (unpublished data). Trypomastigotes separated from the
blood cells by centrifugation were twice washed, diluted to a 3 to
6×10^6 ml concentration and incubated for one hour at $37^{\circ}C$ with

the same volume of human complement or inactivated human serum as the control. The parasites were then counted in a Neubauer chamber and the percentage of lysis determined as follows:

$$\% \text{ lysis} = 100 - \frac{\text{No. parasites incubated with human C} \times 100}{\text{No. parasites incubated with inactivated serum}}$$

FIGURE 2

CURVES OF PARASITAEMIA IN GROUPS OF SPLENECTOMIZED, AND SHAM-SPLENECTOMIZED MICE, INOCULATED WITH 100,000 BLOODSTREAM FORMS OF T. CRUZI Y STRAIN BY SUBCUTANEOUS ROUTE. INOCULATION THREE DAYS AFTER SPLENECTOMY

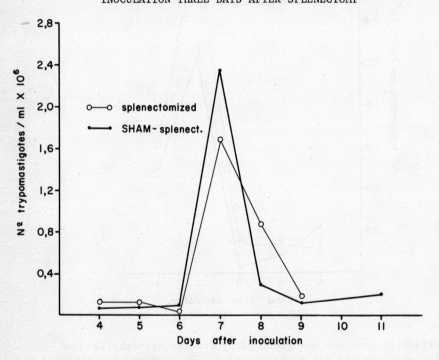

The Table shows the percentage of lysis obtained in experiments using trypomastigotes from splenectomized and sham-splenectomized animals inoculated by intraperitoneal and subcutaneous route. The results show that trypomastigotes from the splenectomized animals are not coated with immunoglobulins as demonstrated by the absence of complement-mediated lysis.

TABLE

PERCENTAGE OF BLOODSTREAM TRYPOMASTIGOTES COLLECTED FROM
SPLENECTOMIZED AND SHAM-SPLENECTOMIZED MICE, LYSED BY ADDITION
OF COMPLEMENT, PARASITES COLLECTED ON THE 8TH DAY OF INFECTION
(POOL OF THREE ANIMALS)

| | Route of inoculation | |
Groups of mice	Subcutaneous	Intraperitoneal
Splenectomized	0	0
SHAM-splenectomized	65	96

The effects of splenectomy in animals in the chronic phase of
Chagas' disease has also been investigated. Groups of mice have
been inoculated with 5×10^4 bloodstream forms from the Y and CL
strains which are highly virulent and kill practically 100% of the
infected animals during the acute phase. Chronic infections were
induced by treating the inoculated mice on the fifth day of
infection with a single dose of 1,000 mg/kg of a nitrofuran
derivative (nifurtimox from Bayer). Prior to splenectomy, both
groups of mice had been twice challenged with 5×10^4 homologous
blood-stream forms and no acute phase developed, an evidence that
a steady host immune response has been brought about by the
original infection.

The strains Y and CL were selected for this experiment on
grounds of their different behaviour in the vertebrate host.
Animals with chronic infection caused by CL strain present a low
but rather constant parasitaemia whereas practically no parasites
are detected in the mice inoculated with the Y strain. Those
differences apparently depend on physiological properties of the
bloodstream forms such as the greater resistance to antibodies
and the diminished uptake by macrophages. A study of the distinct
characteristics of Y and CL populations has been recently
published.(10)

Figure 3 shows the results obtained after splenectomy of
animals chronically infected with Y and CL strains. A small
increase in the number of parasites was detected after splenectomy
in the animals inoculated with the CL strain but the animals have
not developed an acute phase which is characterized by extremely
high parasitaemias similar to those seen in Figs. 1 and 2, reaching

126

FIGURE 3

PARASITAEMIA IN GROUPS OF MICE IN THE CHRONIC PHASE OF
CHAGAS' DISEASE (Y AND CL STRAINS) SUBMITTED TO SPLENECTOMY
AND, AFTER ONE MONTH, A CHALLENGE INFECTION
WITH HOMOLOGOUS BLOODSTREAM FORMS.

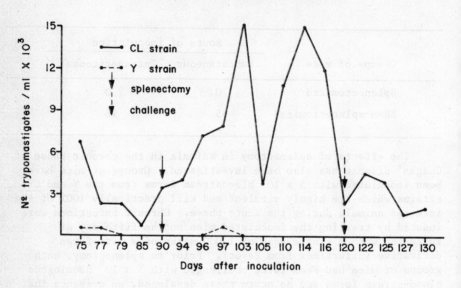

levels as high as 10^7 parasites/ml. The steady sub-patent
parasitaemia observed in the animals inoculated with the Y strain
was not at all affected by splenectomy. The course of infection
was not either affected in the groups of mice chronically
infected with Y and CL strains which had been submitted to a sham-
splenectomy. The indirect immunofluorescence technique using
rabbit anti-mouse immunoglobulin and epimastigote culture forms
was performed before and one month after splenectomy in the
chronic animals. The antibody titres had not been changed by the
spleen removal. A challenge with 100,000 homologous bloodstream
forms was performed in the two groups of splenectomized mice and
no new outbreaks of parasitaemia characteristic of an acute phase
were observed (Fig. 3).

In summary, the experimental evidence demonstrate that, at
least with the Y strain, splenectomy does not play a critical
role in host resistance during the early stages of the infection
despite the importance of the spleen in the production of specific
antibodies able to bind to the parasites. During the chronic

phase the course of infection is practically not affected by the spleen removal. Apparently, this organ is not essential in the protective immunity against T. cruzi.

As we have anticipated, some T. cruzi strains display selective parasitism of cells from the liver, bone marrow and, predominantly, from the spleen. Those populations, therefore, by affecting an organ which is rich in immunologically-competent cells may influence the host immune response. On the other hand, this peculiar interaction with spleen cells, most probably macrophages, may help to interpretate the controversial problem of T. cruzi tissue tropism.

Preferential parasitism of muscle cells or, then, reticuloendothelial, glial or fat cells by T. cruzi has been reported by early authors. Badinez (11) has inoculated dogs and guinea-pigs with different T. cruzi strains isolated from Chile and Brazil, mentioning that some were preferably myotropic and others "reticulotropic". Dogs inoculated with Chilean strains presented a high parasitism of heart and skeletal muscle cells whereas a remarkable low parasitism of cells of the reticuloendothelial system has been observed.(12) Taliaferro and Pizzi (13) studied the parasite distribution and connective tissue reaction to a "reticulotropic" strain of T. cruzi. In general, parasitism was high in the spleen, where parasites were found in both free and inflammatory macrophages as well as in reticular cells but never in lymphocytes; in the liver Kupffer cells and inflammatory macrophages; in skin histiocytes and, finally, in reticular cells of the bone marrow. Parasites were found, "much less often in muscle cells". On the other hand, a clear myotropism was observed in mice inoculated with five T. cruzi strains from Costa Rica; few parasites were detected in the spleen and liver. (14) No parasites were detected in the liver and spleen of mice inoculated with six Mexican strains.(15) Andrade (16) in Brazil, studied 16 strains collected in the same area and described in all of them a proeminent myocardiotropism. Hanson and Roberson (17) described in mice a clear relationship between the number of intracellular stages and the number of circulating bloodstream forms. The higher density of parasites was observed in cardiac, skeletal and smooth muscle but no parasites were again detected in the liver and spleen throughout the period of infection.

More recently, a systematic study of the distribution of intracellular parasites in organs and tissues of mice infected with different T. cruzi strains has been carried out.(10,18) Among the investigated populations, two strains emerged which presented

128

distinct behaviour as regards host-parasite relationship at the cellular level. In the CL strain, a very low parasitism was observed in the spleen, liver and bone marrow; the number of intracellular stages, however, was very high in muscle cells at the late periods of infection. With the Y strain a massive parasitism was present in the spleen red pulp area and marginal zone, Kupffer cells and bone marrow (Figs. 4,5,6). In the spleen

FIGURE 4

NUMBER OF AMASTIGOTES IN HISTOLOGICAL SECTIONS OF THE SPLEEN OF MICE INOCULATED WITH 100,000 BLOODSTREAM FORMS OF T.CRUZI Y AND CL STRAINS (NUMBER OF PARASITES x 10^3 DETERMINED IN 100 MICROSCOPIC FIELDS). (FROM BRENER, 1977).

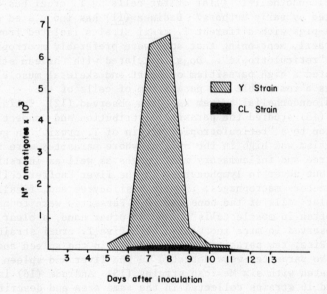

after the 8th day of infection a sudden and marked in situ cell depletion restricted to the red pulp occurs, which parallels the destruction of intracellular parasites (Fig. 7). We are assuming that the selective parasitism of the spleen, liver and bone marrow cells by the Y strain should be considered as a predominent infection of macrophages from the mononuclear phagocytic system (MPS): those organs present high numbers of macrophages; in the spleen, parasites accumulate in the areas rich in macrophages such as the marginal zone and the red pulp but not in the white pulp; in the liver, the parasitism is clearly of the Kupffer cells,

FIGURE 5

HISTOLOGICAL SECTIONS OF THE SPLEEN OF MICE INOCULATED
WITH T. CRUZI. A – INOCULATED WITH Y STRAIN, 7TH
DAY OF INFECTION; PARASITES IN THE RED PULP AREA;
B – INOCULATED WITH THE CL STRAIN, 12TH DAY OF
INFECTION; NO PARASITES ARE OBSERVED. (x 250)

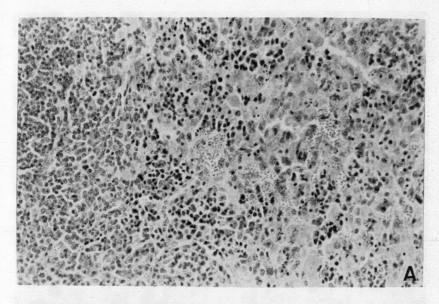

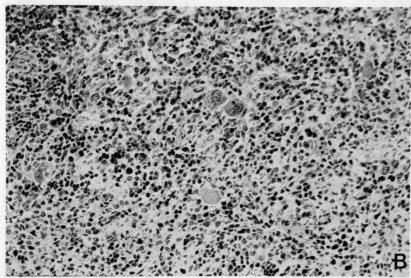

FIGURE 6

INTRACELLULAR STAGES OF T. CRUZI IN THE RED PULP AREA
OF A MOUSE INOCULATED WITH THE Y STRAIN,
7TH DAY OF INFECTION (x 1,000).

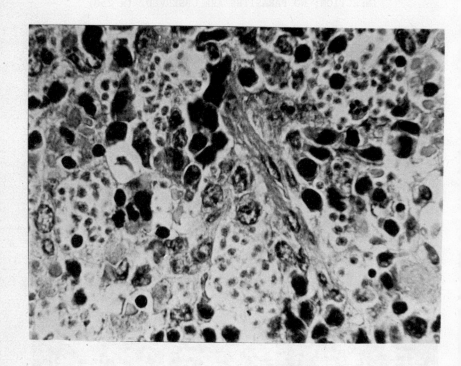

constituents of the MPS; experiments in vitro using mice
peritoneal macrophages and bloodstream forms from Y and CL strains,
closely parallels the findings in the living host, the uptake of
the Y parasites being 20 to 30-fold higher than of CL
trypomastigotes.(Alcantara and Brener, unpublished data)

The results obtained with the CL and Y strains suggest that
they are "polar" types of parasite populations (10) which strongly
differ in their relationship with phagocytic cells from the spleen,
liver and bone marrow. A discussion on this phenomenon should take
into consideration recent concepts on the mechanisms by which

FIGURE 7

CELL DEPLETION IN THE RED PULP AREA OF A MOUSE INOCULATED
WITH T. CRUZI Y STRAIN, 10TH DAY OF INFECTION.

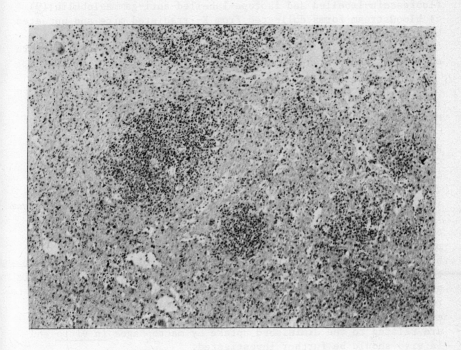

intracellular protozoa enter macrophages and other host cells.(19,
20) The "penetration" of intracellular parasitic protozoa
apparently involves a peculiar sort of endocytosis which seems to
be different from the phagocytosis of inert particles and the
immunological phagocytosis mediated by antibody receptors. In this
kind of "induced phagocytosis" which apparently takes place not
only in macrophages but also in non-professional phagocytic cells,
parasite and cell play active roles in the endocytic process.

The reasons of the selective parasitism as well as the
unsusceptibility of MPS cells to bloodstream forms of different
T. cruzi strains are not yet known. The possibility that they are
related to immunological phagocytosis is not supported by
experimental evidence: a) the tropism for the spleen, liver and

bone marrow is still present in lethally X-irradiated mice which had completely abolished their immune response;(Melo and Brener, unpublished data) b) despite the differences in the infectivity to macrophages, binding of immunoglobulins to the surface of bloodstream forms from both Y and CL strains was confirmed by fluorescein-labelled and isotope-labelled anti-gammaglobulin;(9) c) bloodstream forms collected from X-irradiated mice and uncoated by specific immunoglobulins are not significantly less infective to macrophages than those collected from normal animals.(Alcantara and Brener, unpublished data)

Differences in membrane components or receptors are likely to influence the parasite phagocytosis at the attachment phase. In Plasmodium knowlesi merozoites a coat with the characteristics of a glycoprotein plays an important role in the adhesion of the parasite to the red cell. In T. cruzi bloodstream forms such cytochemical analysis has not yet been performed. Nevertheless, there is much evidence suggesting that membrane peculiarities of the circulating trypomastigotes might account for the differences in the cell-parasite interaction. As mentioned before, bloodstream forms from the Y strain (presenting predominantly slender forms) are strongly agglutinated by specific immune-sera whereas the CL parasites remain unaffected by the antibodies.(3) The Y bloodstream forms are lysed upon addition of human complement but the CL trypomastigotes are able to normally escape from this lytic effect.(9) The behaviour of the CL parasites has been suggested to represent a mechanism of evasion from serum factors. To what extension those peculiarities of the CL strain bloodstream forms are related to the diminished uptake by macrophages in vitro and in vivo should be further investigated.

The macrophage protease-sensitive receptors involved in the uptake of T. cruzi culture forms (20) have not yet been investigated in relation to bloodstream forms. It is difficult, however, at this stage to accept the existence of selective receptors for different T. cruzi populations which would account for the reported differences.

The sudden and extensive destruction of lymphoid cells and parasites in the spleen of animals inoculated with the "macrophagotropic" strains remains obscure. The interpretation of the decline in the number of splenocytes is important since the depletion of immunological competent cells may participate in the known immunosuppression of the acute experimental Chagas' disease. (21) It has been suggested that depletion of T and B lymphocytes from lymph nodes and thymus in mice inoculated with Plasmodium

berghei is involved in the depression of the immune response. (22)
The malaria-induced changes in the thymus and lymph nodes are not
yet clear. The authors suggested that the decrease of B
lymphocytes might be caused by removal of the cells coated with
malarial immune complexes. In the spleen of mice inoculated with
P. berghei, the lymphatic tissue is greatly reduced and the ectopic
erythropoiesis increased. In our material the destruction of
spleen cells as well as parasites is not gradual but rather sudden
and massive, suggesting a relationship between both events. Melo
and Brener (unpublished data) proposed that infected phagocytic
cells in the spleen might serve as target cells for sensitized
lymphocytes; this mechanism has been already described in vitro
by Kuhn and Murnane (18) using spleen cells from infected animals
and parasitized fibroblasts. The fact that cell depletion takes
place in the spleen regions where parasitism is highly predominant -
the red pulp area and the marginal zone - provides further evidence
for this explanation.

The implications of T. cruzi capability of discriminating
among cells in the vertebrate are still speculative. This tropism
for muscle or phagocytic cells seems to be present in experimental
hosts such as dogs (11) in which the infection may resemble the
human disease or even human fetus with congenital Chagas' disease.
(23) Selective parasitism could, then be a more general phenomenon
and participate in the pathogeny of the disease in man.

REFERENCES

1. WEISS, L. The cells and tissues of the immune system. USA,
    Prentice-Hall, Inc., 1972.

2. HANSON, W.L. Immune response and mechanism of resistance in
    Trypanosoma cruzi. In: Chagas' Disease. Pan American Health
    Organization, Scientific Publication, 347: 22 (1977).

3. KRETTLI, A.U. and BRENER, Z. Protective effects of specific
    antibodies in Trypanosoma cruzi infections. Journal of
    Immunology, 116: 755 (1976).

4. NIESCHULZ, O. and WAWO-ROENTOS, F.K. Ueber den Einfluss der
    Milzexstirpation bei Infektionem mit Trypanosoma gambiense
    und Schizotrypanum cruzi. Zeitschrift Immunitätsforschung, 65:
    312 (1930).

5. DIAS, E. Estudos sobre o Schizotrypanum cruzi. Memorias do
    Instituto Oswaldo Cruz, 28: 1 (1934).

6. WOOD, F.D. Experimental studies on Trypanosoma cruzi in California. Proceedings of the Society of Experimental Biology and Medicine, 32: 61 (1934).

7. GALLIARD, H. Infections à Trypanosoma cruzi chez les animaux splenectomisés. Bulletin de la Societé de Pathologie Exotique, 23: 188 (1930).

8. PIZZI, T. and KNIERIM, F. Modificaciones del baso en relacion con la tasa de anticuerpos circulantes en ratones experimentalmente infectados con Trypanosoma cruzi. Boletin Chileno de Parasitologia, 10: 42 (1955).

9. KRETTLI, A.U. and NUSSENZWEIG, R.S. Presence of immunoglobulins on the surface of circulating trypomastigotes of T. cruzi resulting in activation of the alternative pathway of complement and lysis. In: Chagas' Disease. Pan American Health Organization, Scientific Publication, 347: 71 (1977).

10. BRENER, Z. Intraspecific variations in Trypanosoma cruzi: Two types of parasite populations presenting distinct characteristics. In: Chagas' Disease. Pan American Health Organization, Scientific Publication, 347: 11 (1977).

11. BADINEZ, O. Contribucion a la anatomia patologica de la enfermedad de Chagas experimental. Biológica (Chile), 3: 3 (1945).

12. PIZZI, T. Algunos aspectos de la enfermedad de Chagas experimental. Comunicacion preliminar. Biológica (Chile), 3: 53 (1945).

13. TALIAFERRO, W.H. and PIZZI, T. Connective tissue reaction in normal and immunized mice to a reticulotropic strain of Trypanosoma cruzi. Journal of Infectious Diseases, 96: 199 (1955).

14. BICE, D.E. and ZELEDON, R. Comparison of infectivity of Trypanosoma cruzi (Chagas, 1909). Journal of Parasitology, 56: 663 (1970).

15. TAY, J. ET AL. Estudios sobre seis cepas mexicanas de Trypanosoma cruzi. Revista de Investigaciones de Salud Publica (Mexico), 33: 67 (1973).

16. ANDRADE, S.G. Caracterização de cepas do Trypanosoma cruzi isoladas do Recôncavo Baiano. Thesis. Universidade Federal de Bahia, 123 (1973).

17. HANSON, W.L. and ROBERSON, E.L. Density of parasites in various organs and the relation to numbers of trypomastigotes in the blood during acute infection of Trypanosoma cruzi in mice. Journal of Protozoology, 21: 512 (1974).

18. KUHN, R.E. and MURNANE, J.E. Trypanosoma cruzi; immune destruction of parasitized mouse fibroblasts "in vitro". Experimental Parasitology, 41: 66 (1977).

19. JONES, T.C. Macrophages and intracellular parasitism. Journal of the Reticuloendothelial Society, 15: 439 (1974).

20. NOGUEIRA, N. and COHN, Z. Trypanosoma cruzi: mechanism of entry and intracellular fate in mammalian cells. The Journal of Experimental Medicine, 143: 1402 (1976).

21. CLINTON, B.A. ET AL. Trypanosoma cruzi: early immune responses in infected mice. Experimental Parasitology, 37: 417 (1975).

22. KRETTLI, A.U. and NUSSENZWEIG, R.S. Depletion of T and B Lymphocytes during Malarial Infections. Cellular Immunology, 13: 440 (1974).

23. BITTENCOURT, A.L. Congenital Chagas' disease. American Journal of Diseases Children, 130: 97 (1976).

Trypanosoma Cruzi: Cell-mediated Immunity
In Vitro

Nadia Nogueira and Z. Cohn, Department of Cellular Physiology
and Immunology, The Rockefeller University, New York, USA.

In Chagas' disease, acquired resistance does not seem to
depend on antibody alone. Therefore, it is reasonable to consider
the possibility of involvement of cell-mediated immunity, with
specifically reactive lymphocytes acting as mediators. A large
body of evidence suggests that this indeed may be the case.[1]

Through the outstanding work of Mackaness, North and
colleagues,[2,3] we learned that, in dealing with facultative
intracellular bacteria, at least two types of cells were involved
in implementing this type of acquired resistance: activated T
cells, the specific mediators of cellular immunity, and mononuclear
phagocytes, the cells through which microbicidal activity is
expressed. As shown by McGregor,[4] activated T cells, regardless
of specificity, are preferentially attracted to areas of
inflammation, and their contact with antigen is therefore
facilitated. The assembly of mononuclear phagocytes in the area
would ensue as a consequence of the local release of
pharmacologically active molecules, arising from nonspecific
inflammatory events as well as from these antigen-sensitized
lymphocytes. Other products of this specific interaction would in
turn act on macrophages, activating them to a microbicidal state,
and result in phagocytosis and destruction of parasites.

In dealing with obligate intracellular parasites, such as
Trypanosoma cruzi, it is legitimate to postulate that similar
mechanisms may be involved. Within a tissue, foci of "immunological"

inflammatory reactions can be formed around parasitized cells, in much the same way as we observe in tuberculosis, the classical example of this form of immunity. As a result of this reaction, activated macrophages will be able to interiorize and kill any organism released from the parasitized cells.

To confirm this hypothesis we should be able to show, first, that macrophage function is altered during the course of an infection with T cruzi, resulting in cells which are able to kill the parasite. Second, that activated T cells, obtained from animals infected with T cruzi should be able to generate in vitro the factors capable of modifying macrophage function, and that these factors would indeed be able to activate normal macrophages in vitro to a microbicidal state.

INDUCTION OF MICROBICIDAL MACROPHAGES THROUGH IN VIVO INFECTION

The first hypothesis we have previously shown to be true. Our experiments indicate that trypomastigotes of T. cruzi survive quantitatively and replicate in the cytoplasm of both normal, resident mouse peritoneal macrophages, as well as in macrophages induced by the intraperitoneal injection of inflammatory agents. Organisms enter the cells by phagocytosis, are initially enclosed within a phagocytic vacuole and subsequently escape into the cytosol where they replicate.(5) In contrast, during a T. cruzi infection, peritoneal macrophages elicited during a secondary challenge with the specific antigen were capable of destroying 60-75% of the ingested organisms within the first 24 hours. However, by 72 hours, the remaining organisms resumed growth and eventually destroyed the monolayer. Parasite survival and multiplication could be due to heterogeneity in either the macrophage or trypomastigote populations, or else a loss of microbicidal activity of these immune macrophages upon in vitro cultivation.

INDUCTION OF MICROBICIDAL MACROPHAGES IN VITRO BY SPLEEN CELL FACTOR(S)

To investigate the points described above, we needed to understand the conditions for activating macrophages to a microbicidal state, and how to maintain this state once achieved. We focused our attention on the response of normal macrophages in vitro to the products of sensitized lymphocytes, obtained from the spleens of mice infected with T. cruzi. These spleen cell populations were exposed to the specific antigen (heat-killed trypanosomes) for 48 hours, and the cell-free supernatant fluids collected. In previous work, we had demonstrated that these spleen cell factors (SCF) were able to activate normal macrophages in vitro to secrete high levels of a plasminogen activator. T cells were required for the generation of these active supernatant fluids, since pretreatment of the sensitized spleen cells with an anti-θ serum and complement prevent the generation of the factors. We have also been able to induce normal and inflammatory macrophages to display microbicidal activity against T. cruzi by exposure to SCF, in a completely in vitro system,(6) although our initial attempts were unsuccessful.(7) The conditions necessary for such induction were established by using macrophages from a mildly inflamed peritoneal cavity, and are summarized in Fig. 1. Exposure of proteose-peptone (PP) induced

FIGURE 1

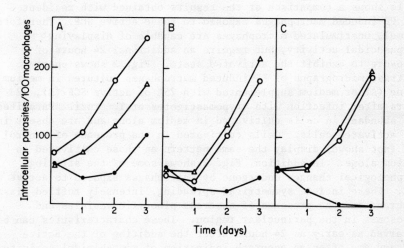

macrophages to SCF from spleen cells of T cruzi infected mice incubated with heat-killed trypanosomes resulted in trypanocidal activity. Pre-incubation of macrophage cultures for 24 hours before infection with active SCF (A-●) resulted in intracellular killing of a portion of the phagocytized inoculum. In the absence of an additional active SCF (days zero-three), organismal growth promptly resumed. Panel B (●) shows the added influence of having the active SCF present not only in the pre-infection period, but added once after infection. Panel C (●) shows that pre-incubation plus the daily addition of active SCF leads to the complete destruction of the intracellular parasites within 48 hours after infection. Control SCF was obtained from cultures of normal spleen cells incubated for 48 hours with the same antigen. Fresh medium alone (Δ) or control SCF (o) were without significant effect. Although not shown in the figure, the % of cells infected and the total cell number were always evaluated. No significant decrease in cell number was observed during the period of the experiments. There were, therefore, two requirements to ensure this state of activation. First, the availability of an active SCF, which was generated by the addition of microbial antigen to spleen cells from a previously infected animal. Second, the daily addition of fresh SCF to the macrophages.

Under optimal conditions revealed in experiments with inflammatory cells, the induction of microbicidal activity in normal, resident macrophages was then promptly demonstrated. The Table shows a comparison of the results obtained with resident and PP-induced macrophages exposed to these active SCF. Therefore, normal, unstimulated macrophages are capable of displaying trypanocidal activity, but require an additional 24 hours of exposure to exhibit the activated state. Fig. 2 shows phase-contrast micrographs of PP-induced macrophages cultured in medium alone (A) or medium supplemented with 25% of active SCF (B), 72 hours after infection with trypomastigotes of T. cruzi. Parasites are abundant in cells cultivated in medium alone and are absent in the activated cells. Cells cultivated in the presence of control SCF (not shown) display the same pattern as those cultivated in medium alone. In addition, Fig. 2 shows some of the striking morphological changes undergone by macrophages exposed to active SCF. These include symmetrical spreading, intensely ruffled plasma membrane and the presence of numerous pinocytic vesicles and lysosomes in the perinuclear region. These characteristics can be observed as early as 24 hours after the addition of the active SCF and are often an accurate indication of microbicidal activity.

SPECIFICITY REQUIREMENTS OF LYMPHOCYTE AND MACROPHAGE FOR INDUCTION OF MICROBICIDAL ACTIVITY

As shown in the Table, the active spleen cell factor nonspecifically activated macrophages to a microbicidal state, since SCF generated from a BCG interaction resulted in microbicidal activity against T. cruzi. The generation of SCF however requires immunological specificity between antigen and sensitized spleen cells.

FIGURE 2

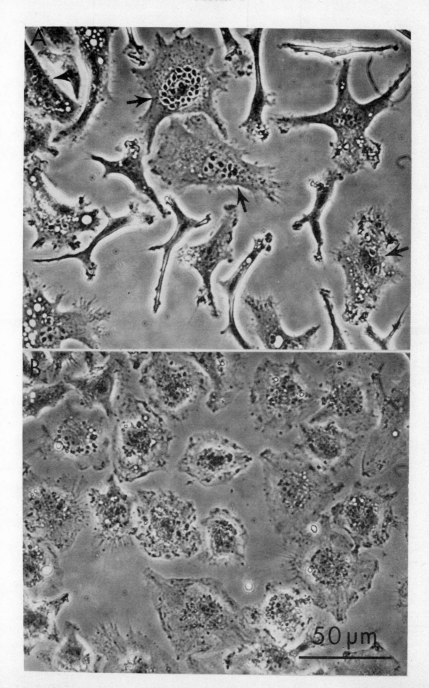

50 μm

TABLE

INDUCTION OF MICROBICIDAL ACTIVITY ON NORMAL AND INFLAMMATORY MACROPHAGES

	Parasites/100 Macrophages				% Infected Macrophages				Macrophages/mm^2	
	3 h	24 h	48 h	72 h	3 h	24 h	48 h	72 h	3 h	72 h
PPalone	55	57	120	253	37	34	39	43	1071	913
PP + BCG SCF	55	10	2	0	38	10	2	0	1170	922
PP + C SCF	54	66	109	172	37	35	34	34	1080	996
Nalone	65	86	192	343	40	47	48	48	1375	1410
N + BCG SCF	63	69	23	0	40	40	18	0	1364	1408
N + C SCF	62	90	42	229	41	42	35	34	1399	1388

The loss of microbicidal activity exhibited by macrophages activated in vivo when kept under in vitro conditions, in the absence of sensitized lymphoid cells, can be restored and maintained by active SCF, suggesting that similar factors may play a role in the intact host. In addition, the nonspecificity of the enhancement of macrophage microbicidal activity is again clearly demonstrated, since the addition of an SCF generated by the addition of PPD to BCG-sensitized spleen cells restores and maintains the trypanocidal activity of macrophages activated in a T. cruzi infected animal   Similarly, the SCF generated through the interaction of heat-killed trypanosomes with T. cruzi sensitized spleen cells leads to enhanced trypanocidal activity of macrophages activated in vivo through BCG infection.

INDUCTION OF MICROBICIDAL MACROPHAGES BY SUPERNATANTS FROM CON A AND LPS-STIMULATED LYMPHOCYTES

Trypanocidal activity could also be induced by supernatant fluids derived from Concanavalin A (Con A) or lipopolysaccharide (LPS) stimulated normal spleen cells. The magnitude of the response obtained with these active SCF was however lower than that produced by antigen-pulsed sensitized lymphocytes. Neither Con A alone (3 µg/ml) or LPS alone (E. coli K235, Phenol-extracted 3 µg/ml, Butanol-extracted 3 and 25 µg/ml) had any effect on the trypanocidal activity of macrophages. These experiments suggest that both bone-marrow and thymus-derived lymphocytes can play a role in the generation of the active SCF.

LOCALIZATION AND POSSIBLE MECHANISMS OF INTRACELLULAR KILLING

Electron micrographs of macrophages activated in vitro by exposure to SCF show partially digested parasites within phagocytic vacuoles 24 hours after infection. At this time phagocytized parasites in control cells have left the confines of the vacuolar system, and are found free in the cytoplasm. For this reason we postulated that the lymphokine would prevent the organisms from escaping from the phagocytic vacuole and therefore allow its digestion within the phagolysosomal system. However, in experiments where the active SCF was added 24 and 48 hours after infection, when all the intracellular organisms were already free in the cytoplasm, trypanocidal activity was also observed. This occurred with a similar time course to that seen when the lymphokines were added before infection. Under these conditions the organisms are found in different stages of digestion within phagolysosomes. This would imply that the organisms are first killed in the cytoplasm and then disposed of by autophagocytosis, or that autophagocytosis would be induced first and death of the organisms would ensue

within the vacuole. Electron micrographs have not as yet
answered these questions. Support for both hypotheses have been
found, that is, morphologically intact organisms within membrane-
bound organelles, or vacuolated parasites seen free in the
cytoplasm.

Since the SCF has no effect on the viability of extracellular
trypomastigotes, we have to assume that it either interacts with
some parasite function only displayed following host cell
infection; it is potentiated by some host cell component, or else
it induces some metabolic event that leads to parasite death.

At this time we do not know the nature of the lymphokine or
the molecular mechanism of the intracellular cytocidal event.
Recent experiments in collaboration with Dr Carl Nathan
(unpublished) have shown a very striking correlation between
hydrogen peroxide release and the expression of microbicidal
activity by in vivo and in vitro activated macrophages. This fact
suggests that hydrogen peroxide may be involved in the
intracellular killing mechanism.

CONCLUDING REMARKS

Utilizing reagents in a completely in vitro environment it
has been possible to induce microbicidal activity of normal
macrophages to kill the obligate intracellular parasite
Trypanosoma cruzi. This was accomplished by exposure of normal
macrophages to a spleen cell factor generated by the addition of
microbial antigen to spleen cells from a previously infected
animal. As in Mackaness' system, the specificity of the system
resides in the lymphocyte population: the generation of the SCF
required immunological specificity between antigen and sensitized
spleen cells. However, once generated, it non-specifically
activated macrophages for intracellular killing.

Our results therefore provide in vitro support for the in vivo
system described by Mackaness and collaborators, extending it to
obligate intracellular parasites. Previous attempts at
reproducing these results under in vitro conditions both in
bacterial as well as protozoan systems (7-12) resulted, under the
best conditions, in only inhibition of growth or short-lived
reduction in viable organisms.

In addition, our results regarding the microbicidal activity
of macrophages adds some more evidence to the fact that there are
major differences in the state of activation achieved by

macrophages exposed to a non-specific inflammatory environment or involved in an "immunological" inflammatory reaction. Some of the so-called "markers" of macrophage activation, described in non-specific inflammatory states, could then be put into the context of a stepwise process, in which rapid spreading, increased ingestion of IgG-coated particles, ingestion mediated by the complement receptor, higher content of lysosomal hydrolases and others, are part of a succession of metabolic and physiological changes associated with the activation process. Macrophage microbicidal activity is often associated with most of the described "activated" functions of macrophages, but the reverse seems not to be true. This would suggest that microbicidal activity against intracellular parasites may be the highest step of a number of metabolic and functional changes that encompass the process of "macrophage activation". Its expression is only observed under conditions in which interaction of antigen with sensitized lymphocytes would lead to a release of lymphocyte products. It is apparent that these lymphokines can induce some of the events mediated by a nonspecific inflammatory reaction, such as spreading enzyme secretion, and others, suggesting that some of these cell-derived products may have the same molecular nature as the humoral mediators of inflammation.

We hope that a better understanding of the molecular nature and site of action of these mediators of cellular immunity will open the way to a rational and efficient approach to immunoprophylaxis and therapy not only of Chagas' disease, but of a large number of parasitic diseases where cell-mediated immunity seems to play a major role in conferring protection.

REFERENCES

1. MAUEL, J. and BEHIN, R. Cell-mediated and humoral immunity to protozoan infections (with special reference to leishmaniasis). Transplantation Reviews, 19: 121 (1974).

2. MACKANESS, G.B. The monocyte in cellular immunity. Seminars on Hematology, 7: 172 (1970).

3. NORTH, R.J. Importance of thymus-derived lymphocytes in cell-mediated immunity to infection. Cellular Immunology, 7: 166 (1973).

4. McGREGOR, D.D. and KOSTER, F.T. The mediators of cellular immunity: IV. Cooperation between lymphocytes and mononuclear phagocytes. Cellular Immunology, 2: 317 (1971).

5. NOGUEIRA, N. and COHN, Z. Trypanosoma cruzi: uptake and intracellular fate in normal and activated cells. American Journal of Tropical Medicine and Hygiene, 26: 194 (1977).

6. NOGUEIRA, N. and COHN, Z. Trypanosoma cruzi: in vitro induction of macrophage microbicidal activity. Journal of Experimental Medicine, 148: 288 (1978).

7. SIMON, H.B. and SHEAGREN, J.N. Cellular immunity in vitro. Journal of Experimental Medicine, 133: 1377 (1971).

8. FOWLES, R.E. ET AL. The enhancement of macrophage bacteriostasis by products of activated lymphocytes. Journal of Experimental Medicine, 138: 952 (1973).

9. HIRSCH, J.G. ET AL. Interactions in vitro between Toxoplasma gondii and mouse cells. Parasites in the immunized host: mechanisms of survival. Ciba Foundation Symposium, 25: 205.

10. ANDERSON, S.E. JR. and REMINGTON, J.S. Effect of normal and activated human macrophages on Toxoplasma gondii. Journal of Experimental Medicine, 139: 1154 (1974).

11. BORGES, J.S. and JOHNSON, W.D. JR. Inhibition of multiplication of Toxoplasma gondii by human monocytes exposed to T lymphocyte products. Journal of Experimental Medicine, 141: 483 (1975).

12. WILLIAMS, D.M. and REMINGTON, J.S. Effect of human monocytes and macrophages on Trypanosoma cruzi. Immunology, 32: 19 (1977).

DISCUSSION

MITCHELL: Can the difference in CL and Y strains be explained by the production of a selective protease by CL which "clips-off" the Fc portion of coating Ig thereby giving a positive for surface Ig using anti-Ig reagents but no agglutination, complement-dependent lysis, or recognition by a phagocytic cell via Fc receptors? Penetration of myoid cells by CL would be an active process of the parasite, entry of Y to phagocytes would be via antibody-dependent phagocytosis. This could be a contributing factor in this very important tissue tropism of various T. cruzi strains.

NOGUEIRA: I do not think that there is any evidence for the production of proteases by trypomastigotes so far. A good possibility to explain the findings with the CL strain is that antibody would be rapidly capped and shed from the surface. The entry of Y strain trypomastigotes into macrophages is not dependent on antibody or complement. We have shown that neither the Fc receptor nor the C3 receptor are involved. We have found that the entry of the Y strain trypomastigotes in non-phagocytic cells is also by phagocytosis and subsequently they also escape from the phagocytic vacuole into the cytoplasm. For this reason, we would expect that the CL strain should behave the same way. What is more likely to be different are the surface components of the two strains, binding to different receptors on the host cells. This difference in receptors would more easily explain the different tropism by the two strains.

FRANK: I would like to add a word of caution about testing for the presence of antibody on parasites by addition of human serum complement and observing lysis. The organism circulates in a sea of active mouse complement. If the organism is lysed by human serum in some cases and not in others this may reflect saturation of sites by the mouse components or differences in antigen steiro specificity which allows human complement to be activated in one case and not another.

NUSSENZWEIG: The finding is that bloodstream forms of certain strains of T. cruzi which have been found to circulate coated with antibodies, are lysed by human complement and, much less efficiently, by mouse complement. Bloodstream forms of other strains, which also induce a rapid antibody response, are not lysed under similar conditions. This is believed to be due to differences in surface antigens and/or differences in the distribution of these antigens. Furthermore there is preliminary evidence indicating that this "bound" immunoglobulin is shed much more rapidly by those parasites which are not lysed by complement (CL strain). Characterization of the surface antigens of these parasites should provide inside into this phenomenon, and possibly also some basis for the clear cut differences of tropisms of these parasites.

BRENER: I fully agree with Professor Nussenzweig that, in order to generalize any concept on the mechanisms of resistance against T. cruzi, we should use more than one T. cruzi population. This can be illustrated by different results obtained with the Y and CL strains as regards in vitro agglutination of bloodstream forms by specific antibodies, passive transfer of resistance by immune sera and in vitro interaction with macrophages. Do you think that we are right on assuming that the spleen parasitism by the Y T. cruzi strain is predominantly by macrophages? What about parasitism of reticular and other kinds of spleen cells?

WEISS: I believe you are right - the distribution and concentration of Y T. cruzi suggests that they are in macrophages. I don't know if reticular cells or other spleen cells are phagocytosed. It's worth a look with the electronmicroscope.

HUMPHREY: It is surprising that macrophages in the spleen white pulp, which are similar to though less numerous than those in the red pulp, escape infection by T. cruzi. Perhaps this reflects the fact that trypanosomes pass directly from the arteries into the red pulp and that there is no mechanism for their return into the white pulp?

COHEN: Is there cross immunity between Y and CL strains in vivo?

BRENER: Yes. If you challenge animals in the chronic phase of Chagas' disease induced by the CL strain with Y strain bloodstream forms they will not display new outbreaks of parasitaemia typical of the acute phase. The same happens with the Y strain.

COHEN: If there is cross immunity between Y and CL strains in vivo, would it be expected that the active supernatant would kill parasites within muscle cells?

NOGUEIRA: This is indeed a possibility. We have been trying to look at the effect of the spleen cell factors in mouse primary myocardial cells and fibroblasts and in some cell lines. However, the spleen cell supernatants are quite toxic to most cells, probably due to lymphotoxin. We will continue to look at that by using more purified factors. It is possible that this lymphotoxin activity plays a role in vivo in tissue destruction. However, humoral immunity may also contribute in conferring protection in vivo.

BATTISTO: Two questions for Dr Nogueira: Your macrophage activating factor has characteristics that closely parallel those of MIF, I'd like to hear your comments relative to this. Secondly, while you have shown that the factor is not directly detrimental to the parasites, are the latter able to interact with the factor and carry it into the macrophage?

NOGUEIRA: At this time we have very little information on the molecular and physicochemical nature of the activating factor. We do not know therefore its relationship to MIF or any other described lymphokine. The experiments in which the factors were added to macrophages 24 to 78 hours after infection indicate that this may not be relevant to the microbicidal effect. In this situation, the factor is added to the macrophages when the parasites are already intracellular, lying in the cell cytoplasm, and the same microbicidal effect is observed.

MITCHELL: It is of some interest that there exists a very close parallelism between the parasiticidal effects of mediators in the in vitro T. cruzi-parasitized macrophage system and observations made by Drs Handman and Burgess at the Hall Institute on the effects of purified colony-stimulating factor (CSF) on in vitro Leishmania tropica-parasitized macrophages. There are several types of central experiments to be performed but the antiparasitic effects of CSF, responsible for promoting haemopoietic colony growth in vitro, are very similar to the effects described by Dr Nogueira.

ZEMBALA: Is anything known about macrophage receptors for the specifically triggered macrophage activating factor(s)? Is the activating factor itself or antigen-factor complex responsible for macrophage activation?

NOGUEIRA: No, we have not looked at that yet. In regard to the second question, we have not looked at that specifically. However, the fact that supernatants from mitogen-induced spleen cells are also able to generate the factor would suggest that such antigen-factor complex is not necessary for macrophage activation.

The Role of the Spleen in Protection Against Murine Babesia Infections

A.C. Allison, J. Christensen, I.A. Clark[1], B.C. Elford,
and E.M. Eugui[2], Division of Cell Pathology, Clinical
Research Centre, Watford Road, Harrow, Middlesex, England.

INTRODUCTION

While there is a considerable body of evidence that the spleen
plays an important role in limiting infections of erythrocytes by
haemoprotozoa, the mechanism by which this protective effect is
achieved is still unknown. The spleen has been considered as a
major source of phagocytic cells,[1-3] antibody-forming cells
[1,3,5,6] and cells required for cellular immunity, a term not
precisely defined.[1,4-6] The importance of removing intraerythrocytic inclusions by pitting has also been emphasized.[7]

Babesias are protozoa, transmitted by ticks, which infect the
erythrocytes of several different mammals. In some tropical and
temperate areas they produce economically important diseases in
cattle (B. argentina, B. bigemina, B. divergens), as well as
infecting dogs (B. canis) and horses (B. equi). An indigenous
infection of wild rodents is produced by B. microti, which has been
transmitted to laboratory rats and mice. The infection in intact
mice is self-limiting, and animals which have recovered from
the infection are resistant to re-challenge for the remainder of
their lifespan. This host-parasite system provides a convenient

---

[1] Present address: Department of Microbiology, Australian
National University, Canberra.

[2] Supported by a research training grant under the Special
Programme for Research and Training in Tropical Diseases, World
Health Organization.

model for investigating the mechanism of recovery from haemoprotozoan infections. Different inbred strains of mice and recombinants can be used to analyse the role of major histocompatibility and other genes in resistance. Transfer of cells from immune donors to sensitive recipients allows studies of cellular immunity. A great deal is known about cellular and humoral immune responses of mice, including lymphocyte subpopulations, and the availability of nude (nu/nu) mice, with a congenital deficiency of mature T lymphocytes, has been useful in analysing the part played by these cells in recovery from infections. Much of the available information on natural killer cells, which will be discussed below, comes from studies of different strains of mice.

It is well known that in cattle splenectomy increases babesia parasitaemias, and this procedure is routinely used in calves to prepare parasites for vaccination.(8) The only described human infections with babesias supposedly of bovine origin have been in splenectomized persons (two in Yugoslavia, one in Ireland and one in California, USA, 9). However, several infections of humans with intact spleens by B. microti in Nantucket Island of the East Coast of Massachusetts, USA, have been recorded. The parasites were isolated in laboratory rodents and identified, and the wild rodent hosts characterized.(9) Hence the absence of a spleen is not a requirement for human infections by babesias, as was once thought. However, it is of interest that many of the cases occurred in old persons, suggesting that their susceptibility to the disease may be increased.

Previous work in this laboratory has established several points about the mechanism by which mice recover from B. microti infections. Since these provide the point of departure for our studies of the role of the spleen in this model, the main conclusions will be summarized:

(1) Mature T lymphocytes are essential for recovery. Nude mice hav parasitaemias that persist for the remainder of their lifespans (10) unless they are reconstituted with syngeneic thymus cells or in ways that will be discussed below.

(2) Serum transferred from recovered mice does not protect recipients, even when given repeatedly in substantial quantities. (11) Hence it is unlikely that the thymus dependence of recovery is due to helper effects on antibody formation.

(3) Mice that have received intravenous or intraperitoneal injections of live Mycobacterium bovis (BCG) (12)or Brucella

abortus, strain 19 (13) or killed Corynebacterium parvum (14) are
strongly protected against B. microti and even against the highly
virulent B. rodhaini. When such mice are injected with very large
doses of parasitized erythrocytes (up to 10^9), they recover within
48 hours from the infection. Even nude mice injected with C. parvum
are resistant to B. microti infections. Hence non-specific
activation of the immune system can substitute for the generation of
a specific immune response during the course of an infection.

(4) The main effect of immunity, as manifested during recovery of
mice from infections with B. microti or in animals protected
non-specifically and challenged with large doses of parasitized
erythrocytes, is degeneration of the parasites within circulating
erythrocytes.(15) The parasites appear shrunken and pyknotic, and
electron microscopic observations show cytoplasmic and nuclear
degeneration, with membrane fusion and the appearance of concentric
lamellae (myelin figures). These degenerate forms of the parasites
rapidly disappear from the circulating blood.

These observations have more recently been extended in our
laboratory, partly with the aim of ascertaining the role of the
spleen in recovery from murine babesia infections. A series of
questions has been asked, and experiments devised in an attempt to
answer them:

(1) Is the spleen necessary for recovery from B. microti infections?

(2) Is the special anatomy of the splenic circulation, which has
been described by Professor Weiss at this meeting, required for
protection against babesia parasites, or will spleen cells perform
the same functions in the absence of a spleen?

(3) Is the spleen required for the generation or manifestation of
non-specific immunity by BCG or C. parvum?

(4) Are spleen cells more efficient than peritoneal or lymph node
cells in transferring immunity to syngeneic recipients?

(5) Which spleen cell subpopulation transfers protective immunity?

(6) Is contact of spleen cells with circulating erythrocytes
required for protection?

(7) How do splenic lymphocytes, macrophages and other cell types
interact in babesia infections?

(8) Can the major differences in susceptibility of different mouse strains to B. microti be associated with any known spleen cell function, such as natural killer activity?

(9) Do spleen cells from immune mice (recovering from babesia infections or non-specifically protected with BCG) produce any factor which could explain the observed degeneration of parasites within circulating erythrocytes?

Some of these questions are easily answered, but most are more difficult to resolve in an unambiguous way. Our work is continuing and the account of it that is given in the next few pages must be read as a progress report rather than a series of definitive statements. It must also be borne in mind that these experiments have been performed with B. microti infections in mice, and that the results are not necessarily applicable to babesia infections of other mammals or to malaria infections. However, a fuller understanding of one host-parasite model may yield information applicable to others.

The spleen facilitates recovery from B. microti infections but is not essential

As shown in Fig. 1, mice which had been splenectomized a week before infection had higher and more prolonged parasitaemias than intact mice, but eventually recovered. When splenectomy was performed at the time of recovery there was a temporary recrudescence of the parasitaemia, followed eventually by complete recovery. These experiments show that the spleen facilitates recovery of mice from B. microti infections, although it is not essential for recovery - as are mature T lymphocytes, for example.

The functional anatomy of the spleen is not required for spleen cells to exert protective effects

Intact CBA mice were given normal or immune spleen cells, infected with B. microti and then splenectomized. The animals receiving immune spleen cells did not develop more than trace parasitaemias,(11) showing that many of the transferred spleen cells migrate to sites other than the spleen and are still able to exert their protective function. Experiments of this kind have the drawback that immunity to B. microti is not sterile, and that small numbers of parasites are carried over with the immune spleen cells. They can multiply in the recipient and actively immunize

FIGURE 1

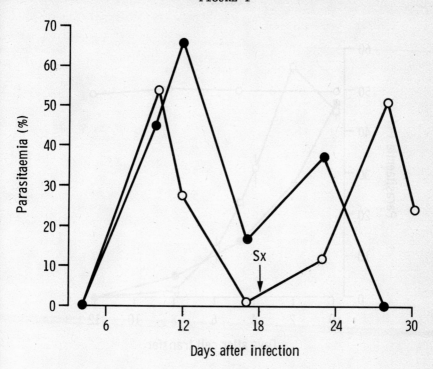

The course of <u>B. microti</u> infections in intact mice (open circles) and following splenectomy at the time of recovery. The course of infections in animals splenectomized before infection is shown by filled circles. Means of six experiments in which the infection was initiated by intraperitoneal inoculation of 10^6 parasitized erythrocytes.

it. For these and other reasons nude mice, which have a persistent high parasitaemia, were given syngeneic immune spleen cells. As shown in Fig. 2, parasitaemias were cleared equally rapidly in nudes that had been splenectomized or were intact.

<u>The spleen is not required for the generation of non-specific immunity by BCG or C. parvum</u>

Mice splenectomized four weeks before intravenous injection of BCG or <u>C. parvum</u> developed full protection against <u>B. microti</u> challenge. When they were challenged with 10^9 <u>B. microti</u> degenerate

FIGURE 2

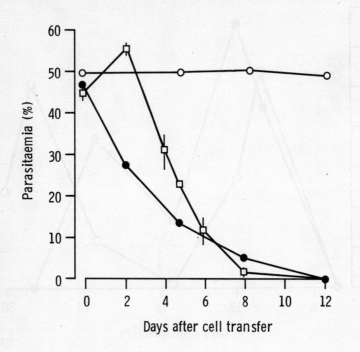

Nude mice carrying B. microti showing persistent parasitaemia (open circles). Clearance of parasitaemia in sham-splenectomized (filled circles) and splenectomized (squares) nude recipients of 10^8 spleen cells from mice that had recovered from B. microti infections two weeks earlier.

forms of the parasites appeared in circulating erythrocytes in the usual way. These persisted much longer in splenectomized than in intact animals (Fig. 3), showing that their removal by pitting or by extraction of erythrocytes containing degenerate parasites from the circulation is much less effective in splenectomized than in intact animals. When mice were splenectomized and infected after such non-specific protection was established, they showed transient parasitaemias. Thus the spleen is not essential for the generation of non-specific immunity, although it is a major site in which such immunity is manifested. In the absence of the spleen other sites ca be non-specifically stimulated by BCG or C. parvum to exert strong protective effects against the parasite.

FIGURE 3

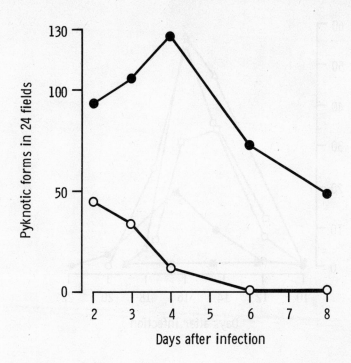

The appearance and disappearance of degenerate forms of
B. microti in the peripheral blood of two groups of mice, one
splenectomized 11 weeks previously (filled circles). The mice had
been given 2×10^7 BCG intravenously 10 weeks previously and
challenged with 10^9 *B. microti*-infected erythrocytes. On each of
the days shown, six uniform fields or smears from each mouse were
searched for pyknotic forms of parasites and the aggregate for each
group of four mice recorded.

Spleen cells are more effective than peritoneal cells or lymph node cells in the adoptive transfer of immunity

As shown in Fig. 4, 10^8 spleen cells from immune donors exerted
highly significant protection on recipients challenged at the same
time, and virtually complete protection when recipients were
challenged two months later. Less protection was seen in recipients
of peritoneal cells recovered by saline wash-out from immune donors,
while animals receiving 10^8 lymph node cells from immune donors were

158

FIGURE 4

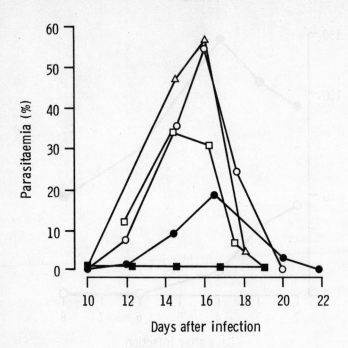

The course of parasitaemias initiated with 10^6 B. microti-infected erythrocytes in CBA recipients of 10^8 normal spleen cells (open circles), 10^8 lymph node cells from immune donors (triangles), 5×10^7 peritoneal cells from immune donors (open squares), 10^8 immune spleen cells (filled circles) and 10^8 immune spleen cells with challenge on day 60 (filled squares). Donors had recovered from B. microti infection 20 days earlier.

not protected at all against B. microti infections. Thus, the cells which confer protection against murine babesia infections are concentrated in the spleen, although in the absence of the spleen strong active immunity can be developed.

Attempts to characterize the spleen cells transferring protective immunity

Spleen cells were treated with anti-Thy 1 serum and complement, with control of the effectiveness of the serum in depleting T lymphocytes. The remaining cells transferred immunity to nude recipients as effectively as the whole population.(11) Enrichment of T lymphocytes by passing the cells over nylon wool columns, with control for depletion of cells with surface membrane immunoglobulins, likewise had no detectable effect on the transfer of immunity to nude recipients.(11) Thus the identity of the spleen cells transferring immunity to babesia infections is still an open question, and will be considered further in the discussion.

Spleen cells can exert their protective effects when introduced into recipients in diffusion chambers

To establish whether direct contact of immune spleen cells with parasitized erythrocytes is required for protection, spleen cells were introduced into the peritoneal cavities of recipients in diffusion chambers limited by 0.8μm pore diameter Millipore filters. These allow exchange of macromolecules, but not cells, with recipients. At the end of each experiment chambers were carefully examined under a stereoscopic microscope to confirm that they were intact.

It was found that nude mice receiving immune spleen cells in diffusion chambers and infected with B. microti cleared their parasitaemias within a month of infection, whereas nude mice receiving diffusion chambers containing normal spleen cells developed the usual persistent high parasitaemia (Fig. 5). These results show that contact of immune spleen cells and parasitized erythrocytes is not required for protection. Presumably a soluble factor produced by the immune spleen cells passes through the Millipore filters and brings about the observed degeneration of parasites within the circulating erythrocytes of the recipient animals. Since this system operates in nude mice, it is unlikely that the soluble factor requires the presence of mature T lymphocytes for its activity. The possible mode of action of a soluble factor secreted by immune spleen cells will be discussed below.

Contact of immune spleen cells and parasitized erythrocytes elicits a granulomatous reaction

In some experiments diffusion chambers containing immune spleen cells and parasitized erythrocytes were placed in intact adult CBA recipients. When these animals were examined one month later, the diffusion chambers were found to be surrounded by a dense mass of cells, which resulted in the adhesion of the chamber to the abdominal wall on the one side and to the spleen on the other.(16) Histological examination (Fig. 6) revealed that the chambers were

160

FIGURE 5

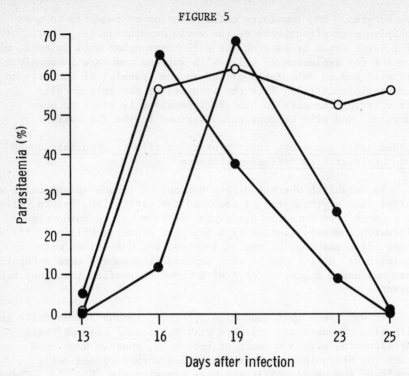

The course of B. microti parasitaemias in nude mice infected
with 500 parasitized erythrocytes on day 0. Nude mice receiving
diffusion chambers containing 10^8 normal spleen cells showed
parasitaemias (open circles) comparable to those in untreated nudes.
Nude mice receiving diffusion chambers containing 10^8 spleen cells
from immune donors developed transient parasitaemias (filled circles)

enveloped by a dense infiltrate of mononuclear cells resembling
macrophages, as well as some delicate blood vessels like those in
granulation tissue. Around the periphery were fibroblasts and a
layer of collagen. It was concluded that immune lymphocytes reacting
with antigen liberated a factor or factors chemotactic for
macrophages, which brought these cells into the vicinity of the
diffusion chambers, as well as migration-inhibitory factor which
immobilizes the infiltrating macrophages, so ensuring their
accumulation. The macrophages themselves liberate factors
stimulating the proliferation of fibroblasts and collagen synthesis.
(16) Since histological examination cannot distinguish macrophages
from other mononuclear cells, such as natural killer cells, the

FIGURE 6

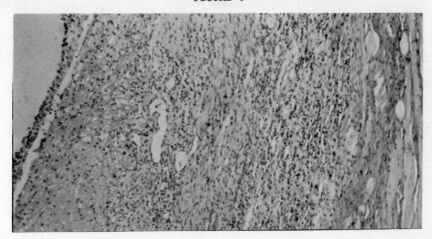

Histological section of the mononuclear cell reaction around a diffusion chamber containing 10^6 immune spleen cells and 10^6 erythrocytes parasitized with B. microti. The diffusion chamber was on the top left. To the right is the muscle of the abdominal wall, to which the infiltrating mass of cells is adherent (x 300).

observations do not exclude the presence of the latter in the infiltrate around the diffusion chambers. Control chambers with immune spleen cells in the absence of antigen or parasitized erythrocytes alone did not show a granulomatous infiltrate.

Differences in the susceptibility of various mouse strains to haemoprotozoan infections

As part of an investigation of the role of histocompability antigens and other inherited factors in the immune response of mice to haemoprotozoan infections, we have followed the course of Babesia microti and Plasmodium chabaudi infections in several strains of mice as well as some recombinant strains. Striking differences in susceptibility have been found (Table I). These are not related to major histocompatibility types but may be correlated with natural killer activity.

B. microti in most strains of mice produces a transient parasitaemia, as shown in Fig. 7 for the BALB/c strain. Similar results have been obtained with the CBA and C57BL, and several other inbred and outbred strains. By contrast in A strain mice the peak

TABLE I

DIFFERENCES IN THE SUSCEPTIBILITY OF VARIOUS MOUSE
STRAINS TO HAEMOPROTOZOAN INFECTIONS

Mouse strain	Haplotype	Persistent B. microti infection	Lethal P. chabaudi infection
A/HeCrc	$H-2^a$	+	+
B10.A/Ola	$H-2^a$	−	−
CBA/CaCrc	$H-2^k$	−	−
C57BL/10ScSACrc	$H-2^b$	−	−
BALB/cCrc	$H-2^d$	−	−

parasitaemia is higher than in other strains and the mice do
not recover from the infections. They have patent parasitaemias
that persist for several months (in some more than six months)
after which the animals die, possibly as a result of complications
due to the persistent infections. In some A mice the level of
parasitaemia fluctuates, with repeated peaks of moderately high
parasitaemia separated by periods of low parasitaemia. B10.A mice,
which have the same major histocompatibility type as the A mice,
recover from the infections after a slightly longer interval than
in BALB/c or C57BL mice.

P. chabaudi likewise produces moderate transient parasitaemias
in most strains of mice. However, in strain A mice the peak
parasitaemias are higher than in other strains and the animals always
die (Fig. 8). Again the B10.A mice, which have the same haplotype
as the A mice, recover from P. chabaudi infections.

The persistent B. microti and lethal P. chabaudi infections in
strain A mice resemble the babesia and malaria infections previously
observed in this laboratory in nu/nu mice, which have a congenital
deficiency of mature T lymphocytes.(10) Since mice of strain A
have normal numbers and functions of T lymphocytes, the reason for
their susceptibility to haemoprotozoan infections must be different.
A property of A strain mice which has emerged recently is the low
capacity of their spleen cells to kill sensitive target cells
(natural killer or NK activity) as compared with other strains.(17,
18) The possible role of NK cells in immunity to haemoprotozoa is
considered further in the discussion.

FIGURE 7

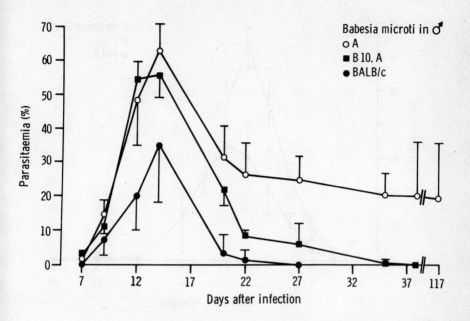

B. microti parasitaemias initiated with 10^6 infected erythrocytes in young adult male mice of strain A (open circles), strain B10.A (squares) and BALB/c (filled circles). Mice of strain A do not recover from the infection.

Increased membrane transport in parasitized red cells

It has been known for several years that the permeability of erythrocytes changes after infection with intracellular parasites. (19,20) Our observations have shown that this is not a general effect and that the parasite selectively alters the membrane transport characteristics of its host cell. We have found that in mouse red cells infected with B. microti relatively little change occurs in the rate of influx of phosphate, D-glucose or inosine (Fig. 9); by contrast, the influx of glycine increases from 6.8 ± 0.8 to 77.2 ± 12.9 (± 1 s.e.,n = 18) μmoles/g cells.hr. The greatest change occurs in the influx of L-glucose, which increases from a mean value of 0.30 ± 0.01 to 25.5 ± 1.6 (± 1 s.e., n = 17) μmoles/g cells.hr. L-glucose is a useful marker molecule since it

164

FIGURE 8

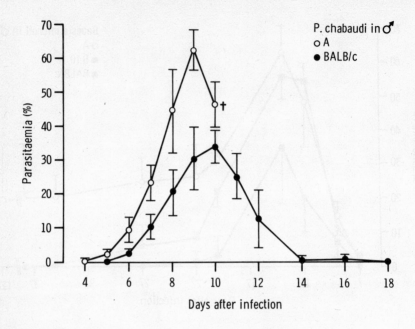

P. chabaudi parasitaemias initiated with 10^6 infected erythrocytes in young male mice of strain A (open circles) and BALB/c (filled circles). All A mice are dead by day 11 whereas all BALB/c mice recover.

is not normally present in red cells and is not metabolized by the red cell or the parasite, so that changes in pool size or intracellular consumption do not affect rates of uptake. Also from Table II it can be seen that the permeability of parasitized cells to L-glucose increases while the parasitaemia is still rising.

Of greater interest from a physiological viewpoint is the increased uptake of myo-inositol that occurs in infected erythrocytes (Fig. 10). An anlysis of such uptake curves showed that the influx of inositol is independent of the parasitaemia (Fig. 11) and that on average the influx is increased in infected cells by a factor of 50 from 0.0030 ± 0.0006 to 0.15 ± 0.03 ($\pm$ 1 s.e., n = 4) μmoles/g cells.hr.

FIGURE 9

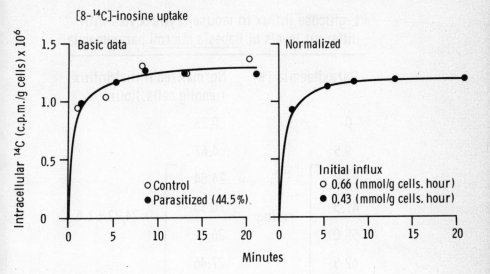

^{14}C-inosine uptake into parasitized (●) and uninfected (o) mouse red cells was measured in a conventional way using tritiated inulin as an extracellular marker. Each uptake curve was numerically fitted to a reciprocal function of the form $F(t) = A - \left(\dfrac{Ac}{c + t} \right)$ where t is time and A and c are constants given by the least squares fit to the data. From a count of the parasitaemia, a correction was made for the contribution from the uptake into uninfected cells in the parasitized group; this corrected curve was then normalized to 100% parasitaemia and a further reciprocal fit carried out. The ratio A/c gave the initial slope of each curve at time zero, from which an influx value was calculated using a measured value of extracellular specific activity. In some experiments where the uptake was slow, linear regression analysis was used instead. In all experiments the TCA-soluble fraction of the uptake was sampled. Further details of the experimental protocol will be published elsewhere.

TABLE II

L-glucose influx in mouse erythrocytes with different levels of Babesia microti parasitaemia

Parasitaemia (%)	Normalized initial influx (μmol/g cells. hour)
0	0.23
9.5	4.67
22.8	24.84
47.5	20.35
58.0	26.41
62.5	27.46
67.5	25.82
46.3 falling	9.82

rising (22.8–62.5) = 24.77 ± 1.6

Effects of spleen cell extracts on permeability of infected erythrocytes

When infected red cells are injected into an immune animal, intracellular degeneration of the parasite occurs rapidly. Our electron microscopic observations (15) have shown that the parasites become shrunken, with fusion of intracellular membranes and development of myelin figures. They resemble the autophagocytosis and other changes seen in starved protozoa such as Amoeba and Tetrahymena.(21) A possible explanation is that the effector cells in immunity release a factor that binds to transport sites on the membrane of infected erythrocytes in such a way that they block the influx of a substance that is required for the normal developmer of intracellular parasites.

FIGURE 10

Myo- [2-³H]-inositol uptake
Increasing parasitaemia

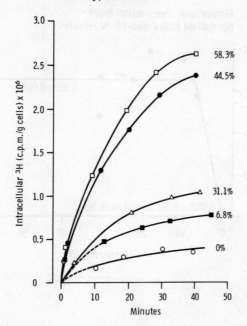

Uptake of $\left[{}^{3}H\right]$-myo-inositol into red cells with increasing parasitaemias. Hydroxy $\left[{}^{14}C\right]$-methyl inulin was used as the extracellular marker.

To test this hypothesis we have investigated membrane transport in the erythrocytes of animals recovering from babesia infections. Moreover, extracts have been prepared from sonicates of spleen cells from normal animals, animals recently recovered from infections and animals made non-specifically immune by BCG. These extracts and control materials have been incubated with babesia-infected erythrocytes and then the rates of influx of L-glucose, glycine and inositol have been measured.

As shown in Table II, the rates of transport of L-glucose are greatly increased during the phase of rising parasitaemia. During the phase of falling parasitaemia the rate of transport is significantly less than in the phase of rising parasitaemia (despite

FIGURE 11

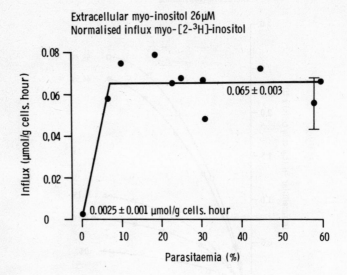

Extracellular myo-inositol 26μM
Normalised influx myo-[2-³H]-inositol

Normalized influx data obtained from an analysis of the uptake of labelled myo-inositol into red cells with varying parasitaemias. Some of the basic influx curves are shown in Fig. 10. For the sake of clarity standard error bars have been shown only on one point.

the presence of reticulocytes, which have increased permeability compared with mature red cells). However, the rate does not fall to the extremely low level in normal erythrocytes.

Figure 12 shows the inhibition of uptake of glycine into parasitized mouse erythrocytes produced by an extract from the spleen of a BCG-treated animal. Similar results have been obtained showing inhibition of the uptake of both L-glucose and myo-inositol by extracts of activated spleen cells. Sometimes (in about one experiment in five) extracts of normal spleen cells are likewise inhibitory, possibly because they have been activated by a recent infection.

169

FIGURE 12

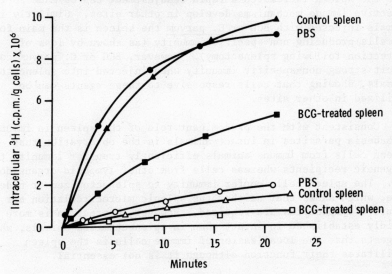

Inhibition of $\left[^3H\right]$-glycine influx into red cells with spleen cell extracts prepared from a BCG-treated animal (2 x 10⁷ viable units BCG at least three weeks previously). Red cells were suspended at a haematocrit of 8% and the extracts from sonicating 7 x 10⁷ spleen cells were included in the basic suspending medium (phosphate-buffered saline PBS). Cells were incubated at 37°C for 30 min before the start of the uptake of $\left[^3H\right]$-glycine. Carrier glycine (1 mM) was present in all suspensions. Initial influxes (μmoles/g cells.hr) were (o) 106, (Δ) 71, (■) 24 for the parasitized groups of cells and (o) 7.2, (Δ) 2.8, (□) 2.2 for uninfected red cells in PBS, control spleen cell and BCG-treated spleen cell supernatants respectively.

DISCUSSION

Does any clear picture of the role of the spleen in immunity to babesia infections emerge from these observations? Certain points seem firmly established while others are still uncertain. The spleen facilitates recovery from babesia infections but is not essential. In splenectomized animals recovery takes longer than in intact animals, but immunity to challenge is solid in both groups. Splenectomy after recovery results in a transient burst of parasitaemia. This suggests that in intact animals the cells suppressing parasitaemia are concentrated in the spleen, but that when the spleen is removed a rapid readjustment can be made so that effective immune mechanisms develop in other sites. Similarly in animals injected with BCG or C. parvum the spleen is the main focus of cells producing non-specific immunity (as shown by loss of protection following splenectomy). However, BCG or C. parvum can elicit strong non-specific immunity when injected into splenectomize animals, showing that cells responsive to these agents can be mobilized in other sites.

Consistent with the predominant role of the spleen in immunity to babesia parasites in intact animals is the observation that spleen cells from immune animals efficiently transfer immunity to syngeneic recipients whereas cells from other lymphoid organs do not. The spleen cells confer immunity to splenectomized or nude mice, which shows that neither the splenic microcirculation nor mature T lymphocytes are required. However, protection is more rapidly established in intact than in splenectomized animals, which suggests that the localization of immune cells in the spleen facilitates their function although it is not essential.

In immune animals or those receiving spleen cells from immune donors, the main observed effect is degeneration of parasites within circulating erythrocytes. In splenectomized animals passively protected with spleen cells the degenerate forms continue to circulate for much longer than in intact animals. This is evidence that the splenic microcirculation is required for the removal of parasite remnants by pitting or clearance of cells containing degenerate parasites from the circulation (it is not yet certain which of these occurs). This is consistent with the classical view (see 7) that the spleen is required for removal of Howell-Jolly bodies, Heinz bodies and other erythrocytic inclusion, although as far as we are aware it has not previously been established that this is a function of the microcirculation rather than of spleen cells themselves. Indeed, the main special function of the splenic microcirculation seems to be removal of inclusions by pitting or trapping of cells with decreased deformability such as spherocytes. This function is evidently not essential for immunity to babesia parasites, although it makes some contribution.

Spleen cells transfer immunity to babesia infections more efficiently than lymph-node cells. This is one of the few clear differences to emerge between spleen cells and the cells of other lymphoid organs. The difference may be interesting or trivial. It is interesting if a special subpopulation of cells, present in much higher concentration in the spleen than in lymph nodes, is required for immunity. It is trivial if antigens from parasites in the circulating blood elicit an immune response in spleen cells whereas lymph node cells are optimally stimulated by antigens in afferent lymph. This could be checked by immunizing mice subcutaneously with parasitized erythrocytes and looking for the capacity of cells in draining lymph nodes to transfer immunity.

In animals inoculated with babesia parasites there is marked splenomegaly at the time when they are recovering from the infection. Mature T lymphocytes are required for recovery, and the splenomegaly following infection of nude mice is much less than in intact animals. This suggests that a thymus-dependent immune response to parasite antigens is important for splenomegaly. The experiments described in this chapter, in which immune spleen cells and parasitized erythrocytes placed in diffusion chambers in the peritoneal cavity elicited a dense mononuclear cell infiltrate, provide evidence for the recruitment of these cells from the circulation. Wyler (22) has described the presence in the spleens of mice recovering from malaria infections of a mononuclear chemotactic factor, and this, together with macrophage migration-inhibitory factor, could explain the accumulation of mononuclear phagocytes in the spleens of infected animals. Possibly formation of macrophages from precursors in the spleen, a haemopoietic organ in the rodent, contributes to the splenomegaly in parasitic infections. But proliferating lymphocytes and macrophages are not the only cells accumulating in the spleens of animals responding to haemoprotozoan infections. Natural killer cells will be discussed below, and other cell types may also be involved.

The subpopulation of spleen cells transferring immunity is still uncertain. The cells were not affected by treatment with anti-Thy 1 serum and complement, which suggests that they are not T lymphocytes. They were not depleted by passage through nylon wool columns, which removes many B lymphocytes and macrophages. Clearly the characterization of these cells should be a high-priority aim for future research. It should also be established whether they are effector cells or activate effector cells in recipients.

How do the immune spleen cells exert their effects? It seems unlikely that the formation of antibodies preventing penetration of merozoites into erythrocytes is important. Transfer of serum does not protect. Although antigenic variation does occur in babesias,

(23) it is not a major factor in escape from immunity, as it is in trypanosomes and some species of malaria parasites. Mice that have recovered from B. microti infections are immune to all variants of this parasite that have been tested and even to B. rodhaini. The observation that parasites degenerate within circulating erythrocytes during the recovery period implies that the parasites have penetrated into erythrocytes. It implies also that opsonization of parasitized erythrocytes for phagocytosis can only be a contributory factor in immunity: otherwise cells with degenerating parasites would not continue to circulate. The same observation makes it improbable that lysis of parasites or infected erythrocytes by immune T lymphocytes or by an antibody-dependent cell-mediated mechanism is a major protective factor, and our attempts to demonstrate such lytic effector activity have been uniformly negative.

Spleen cells from immune donors placed in diffusion chambers in the peritoneal cavities of nude recipients clear their parasitaemias. Hence contact between the immune spleen cells and circulating parasitized erythrocytes is not required for termination of the infection. We are forced to the conclusion that the final effector mechanism is the secretion by cells in the spleens of animals specifically immune, or made non-specifically resistant by BCG or some other agent, of a factor which inhibits the development of parasites in erythrocytes. The final effector mechanism probably does not involve mature T lymphocytes since nude mice can be protected by inoculation of C. parvum.

What cells might elaborate such a protective factor and how might it work? The two most obvious cell types for consideration are macrophages, which are now known to release a variety of secretory products (24) and natural killer (NK) cells. Our attempts to abrogate immunity with silica have been negative.(11) Although the use of silica to eliminate effects of macrophages in vivo is limited because of problems with dosage, timing and access to susceptible cells,(25) this provides an indication that macrophages may not be the main cell type involved.

An alternative possibility, that NK cells might play a role in haemoprotozoan infections, is supported by the observation that mice of strain A are clearly more susceptible than other strains. Strain A mice are unable to recover from infections with B. microti or B. chabaudi. They also have much lower NK activity in their spleens (measured by capacity to lyse tumour target cells) than the other strains tested. Of course the correlation may be fortuitous, and other strains with low NK activity must be studied, as well as hybrids, back-crosses and bone-marrow chimaeras, to ascertain whether low NK activity and susceptibility to haemoprotozoan infections segregate together or independently. It may well be relevant that intravenous or intraperitoneal injections of BCG or C. parvum considerably augment NK activity of spleen or peritoneal cells (26,27) and increase the resistance of mice to haemoprotozoan infections.(12,14) Conversely, corticosteroids, which increase susceptibility to haemoprotozoan infections and eliminate resistance produced by BCG, (12) depress NK activity.(28)

It is also of interest that interferon, which is formed during the course of haemoprotozoan infections (29) increases NK activity, (30) and interferon is apparently an intermediate in the activation of NK cells by a variety of agents, as shown by inhibition with anti-interferon serum. Activated T lymphocytes also release interferon, which can increase the virucidal activity of macrophages (31) and may likewise increase NK activity.

On the basis of these observations a working hypothesis can be suggested (Fig. 13). Recovery from babesia infections depends on a specific immune response mediated by T lymphocytes. Mice deprived of mature T lymphocytes cannot recover.(10) Mice infected with Plasmodium yoelii show marked T cell stimulation,(32) and the same is probably true of mice infected with B. microti. The responding T lymphocytes secrete products that recruit and activate macrophages, as well as interferon.(31) Macrophages can be activated in an immunologically specific manner by products of T lymphocytes or non-specifically by BCG, C. parvum, Brucella abortus and other agents. Among the products liberated by activated macrophages is interferon, which increases NK activity. Macrophages, NK cells or cells of some other type secrete a factor or factors that inhibit the development of parasites in erythrocytes.

The mechanism by which such factors exert their inhibitory effects is still a matter of speculation. However, their activity may be related to transport systems in the membrane of the infected red cell. We have shown that transport of certain substances is selectively increased in red blood cells infected with haemoprotozoa. Some of these (e.g. L-glucose) are useful as markers for studies of membrane permeability. Others may be required for the metabolism of the parasite. For example, inositol phosphatides are found in many parasitic protozoa, including haemoprotozoa, and phosphatidylinositol occurs in much greater amounts than in mammalian cells.(33) The turnover of polyphosphoinositides in parasitic protozoa is very rapid. The concentration of inositol derivatives in erythrocytes is very low, which is consistent with the

174

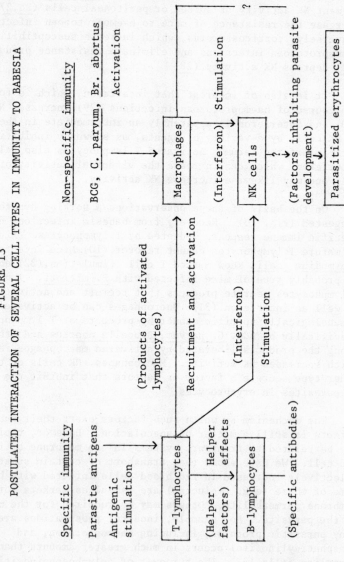

FIGURE 13

POSTULATED INTERACTION OF SEVERAL CELL TYPES IN IMMUNITY TO BABESIA

Cell types are shown in boxes and products in brackets and effects are indicated with arrows

low rate of transport into uninfected cells, whereas the greatly increased transport in the infected cells could be a specific activation by the parasite of a transport system to meet a growth requirement. Hence interference with the supply of inositol from extracellular sources could limit the development of the parasites.

We have found that during the phase of recovery from a babesia infection the rate of transport into infected cells is less than during the phase of rising parasitaemia. Extracts of cells from the spleens of recovered animals or animals made non-specifically immune by BCG inhibit transport. Thus the parasite specifically activates certain transport systems, and factors in immune spleen cells inhibit them. Whether this inhibition is sufficient to explain the observed degeneration of parasites within cells remains to be established.

REFERENCES

1. TALIAFERRO, W.H. Functions of the spleen in immunity. American Journal of Tropical Medicine and Hygiene, 5: 391 (1956).

2. JANDL, J.H. ET AL. The destruction of red cells by antibodies in man. I. Observations on the sequestration and lysis of red blood cells altered by immune mechanisms. Journal of Clinical Investigation, 36: 1428 (1957).

3. CROSBY, W.H. Normal functions of the spleen relative to red blood cells. Blood, 14: 399 (1959).

4. ROBERTS, J.A. Adoptive transfer of immunity to Babesia rodhaini by spleen cells from immune rats. Australian Journal of Biology and Medical Science, 46: 807 (1968).

5. STECHSCHULTE, D.J. Cell-mediated immunity in rats infected with Plasmodium berghei. Military Medicine, 134: 1147 (1969).

6. PHILLIPS, R.S. Plasmodium berghei: Passive transfer of immunity by antiserum and cells. Experimental Parasitology, 27: 479 (1970).

7. CROSBY, W.H. Siderocytes and the spleen. Blood, 12: 165 (1957).

8. CALLOW, L.L. and TAMMEMAGI, L. Vaccination against bovine babesiosis. Australian Veterinary Journal, 43: 249 (1967).

9. HEALY, G.R. ET AL. Human babesiosis: reservoir of infection on Nantucket island. Science, 192: 479 (1976).

10. CLARK, I.A. and ALLISON, A.C. Babesia microti and Plasmodium berghei yoelii infections in nude mice. Nature, 252: 328 (1974).

11. CLARK, I.A. Immunity to intra-erythrocytic protozoa in mice, with special reference to Babesia sp. Ph.D. thesis, University of London (1976).

12. CLARK, I.A. ET AL. Suppression of babesiosis in BCG-infected mice and its correlation with tumour inhibition. Infection and Immunity, 17: 430 (1977).

13. HEROD, E. ET AL. Protection of mice against haemoprotozoan Babesia microti with Brucella abortus strain 19. Clinical and Experimental Immunology, 31: 518 (1978).

14. CLARK, I.A. ET AL. Protection of mice against Babesia spp. and Plasmodium spp. with killed Corynebacterium parvum. Parasitology 74: 9 (1977).

15. CLARK, I.A. Intra-erythrocytic death of the parasite in mice recovering from infection with Babesia microti. Parasitology, 75: 189 (1977).

16. ALLISON, A.C. ET AL. Cellular interactions in fibrogenesis. Annals of the Rheumatic Diseases, 36: 8 (1977).

17. KIESSLING, R. ET AL. Genetic variation of in vitro cytolytic activity and in vivo rejection potential of non-immunized semi-syngeneic mice against a mouse lymphoma line. International Journal of Cancer, 15: 933 (1975).

18. HERBERMAN, R.B. ET AL. Natural cytotoxic reactivity of mouse lymphoid cells against syngeneic and allogeneic tumours. I. Distribution of reactivity and specificity. International Journal of Cancer, 16: 216 (1975).

19. SHERMAN, I.W. and TANIGOSHI, L. Glucose transport in the malarial (P. lophurae) infected erythrocyte. Journal of Protozoology, 21: 603 (1974).

20. HOMEWOOD, C.A. ET AL. Permeability of erythrocytes from mice infected with Babesia rodhaini. Annals of Tropical Medicine and Parasitology, 69: 429 (1975).

21. DE DUVE, C. and WATTIAUX, R. Functions of lysosomes. Annual Review of Physiology, 28: 435 (1966).

22. WYLER, D. and GALLEN, J.I. Spleen-derived mononuclear cell chemotactic factor in malarial infections: a possible mechanism for splenic macrophage accumulation. Journal of Immunology, 118: 478 (1975).

23. BROWN, K.N. Antigenic variation and immunity to malaria. Amsterdam ASP (Elsevier, Excerpta Medica, North Holland) Ciba Foundation Symposium No. 25, New Series (1974).

24. ALLISON, A.C. Macrophage activation and nonspecific immunity. International Review of Experimental Pathology, 18: 303 (1978).

25. ALLISON, A.C. Fluorescence microscopy of lymphocytes and mononuclear phagocytes and the use of silica to eliminate the latter. In: In Vitro Methods of Cell-Mediated and Tumour Immunity (Bloom, B.S. and David, J.R. eds.) Academic Press, New York, p.395 (1976).

26. OJO, E. ET AL. An analysis of conditions allowing Corynebacterium parvum to cause either augmentation or inhibition of natural killer cell activity against tumour cells in mice. International Journal of Cancer, 21: 444 (1978).

27. HERBERMAN, R.B. ET AL. Augmentation of natural cytotoxic reactivity of mouse lymphoid cells against syngeneic and allogeneic target cells. International Journal of Cancer, 19: 555 (1977).

28. HOCHMAN, P. and CUDKOWICZ, G. Different sensitivities to hydrocortisone of natural killer cell activity and hybrid resistance to parental marrow grafts. Journal of Immunology, 119: 2013 (1977).

29. HUANG, K.Y. ET AL. Interferon induced by Plasmodium berghei. Science, 162: 123 (1968).

30. GIDLUND, M. ET AL. Enhanced NK activity in mice injected with interferon and interferon inducers. Nature, Lond. 273: 759 (1978).

31. VIRELIZIER, J.L. ET AL. Production by mixed lymphocyte cultures of a type 11 interferon able to protect macrophages against virus infections. Infection and Immunity, 17: 282 (1977).

32. JAYAWARDENA, A.N. ET AL. T-cell activation in murine malaria. Nature, Lond. 258: 149 (1975).

33. HOLZ, G.G. Jr. Lipids and the malaria parasite. Bulletin of the World Health Organization, 55: 237 (1977).

DISCUSSION

MILLER: You mentioned that A strain mice behave as nude mice in their susceptibility to Babesia. As you stated, A strain mice have very low NK cell levels. However, nude mice do have marked NK cell activity. What then would be the role of NK cells in resistance in this system? Could T-cell dependent interferon production play a role in resistance via NK cell activation?

ALLISON: Interferon increases NK activity and stimulated T lymphocytes produce interferon, so this is an attractive hypothesis linking the requirements of mature T lymphocytes for recovery from babesia infections and the postulated role of NK activity in this process.

SPRENT: Don't nude mice have increased NK activity?

ALLISON: The NK activity of nude mice housed under conventional conditions (where they may be stimulated by lipopolysaccharide and other bacterial products) is slightly increased over that of intact mice. This activity is, however, much less than in the spleens of animals that have been injected with BCG.

TRAGER: The nutritional effect is naturally of great interest to me. There are old observations suggesting similar effects in malaria. Thus Taliaferro described the "crisis forms" and, even more interesting, noted that as an infection progressed, the average number of merozoites per parasite decreased. This suggests that their nutrition was less good. In work with the bird malaria Plasmodium lophurae I observed that injection of plasma from older resistant birds into young infected ones again produced a decrease in the average number of merozoites per parasite.

FRANK: You say that when the spleen is removed from an animal who has controlled his infection there is a brief return of parasitaemia. How long after the initial infection does this effect persist? Do you have any explanation of where or how the organisms persist?

ALLISON: Mice infected with Babesia microti seldom, if ever, show sterile immunity. The ongoing immune response suppresses the parasitaemia to undetectable levels. Presumably small numbers of parasites persist in erythrocytes in the circulation or secluded in the bone marrow or other sites. The site of persistence is not the spleen since parasitaemia recurs after splenectomy.

MORRIS: You showed that spleen cells from previously infected animals protected and so concluded, I think, that the spleen as a whole was not involved in the partial protection. Do non-immune or virgin spleen cells protect, and if not, are you justified in drawing the above conclusion? For once a clone of sensitized cells is available then the spleen may be irrelevant, just as in the renal allograft model I mentioned previously.

ALLISON: Virgin spleen cells injected into a normal or splenectomized recipient do not protect against babesia infection. It is an attractive hypothesis that the anatomical arrangement of the spleen favours primary responses but is less important for secondary, just as germinal centres favour the development of memory cells. However, the presence of the spleen is not essential for a primary immune response, since splenectomized mice recover from infections with B. microti after higher and more prolonged parasitaemias than in intact parasitaemias.

LANGHORNE: Does BCG protect A strain mice against B. microti? If so, do these mice then have enhanced NK activity?

ALLISON: Mice treated with BCG certainly show increased NK activity. We have not yet attempted to protect A strain mice with BCG.

SPRENT: Two questions. Firstly, do purified T cells from immune spleens confer protection adoptively? Secondly, do lymph nodes become protective at late stages post-immunization?

ALLISON: Treatment of spleen cells with anti-Thy 1 sera does not abolish their capacity to protect recipients. We do not know whether lymph-node lymphocytes become protective in the late stages of infection.

TRAGER: What were conditions for testing effect of spleen supernate on uptake? How much was the inositol uptake decreased by the spleen supernate?

ALLISON: Infected cells were incubated one hour with the spleen supernate. About 10 fold.

CLAYTON: When you did your experiments with spleen cell supernatants, did you control for the possibility that the supernatants you added might have contained cold substrate?

ALLISON: Yes, our dilution experiments eliminate this possibility, and some of the substrate used, such as L-glucose, were not present in the spleen cell supernatants.

CLAYTON: With reference to your permeability experiments, could you comment on the change in transport of L-glucose as this is unlikely to be present physiologically? Also, to what extent are the permeability changes in parasitized cells a consequence of parasite metabolism? This could affect transport rates by changing a) concentration gradients by utilization of substrates for macromolecular liosynthesis and b) efflux of products of parasite metabolism which could be coupled to influx of other molecules.

ALLISON: The uninfected erythrocyte is nearly impermeable to L-glucose, unlike D-glucose. After infection, influx of L-glucose increases more than a hundredfold. This is not because of a concentration gradient, which is the same in infected and uninfecte cells. The parasite does not utilize L-glucose for its metabolism. We suppose that the L-glucose transport site is normally used for transporting something else, but the competition experiments so far carried out have not identified the agent. The main point is that the changes in permeability in the infected erythrocytes are selective.

McGREGOR: The outbreak of human babesiosis in Nantucket has one interesting feature - almost all the infections have occurred in ol subjects. Would you care to comment on possible differences in spleen from young and old persons?

ALLISON: In mice NK activity declines in old animals. If the same is true of old humans there is a ready explanation for the susceptibility to babesia infections. However, other changes in the immune system of old persons may be involved.

MORRIS: Would you explain the failure of lymph-node cells to protect on the basis of lack of natural killer cell activity?

ALLISON: Certainly NK activity is less in lymph nodes than in spleen. Whether this explains the difference in efficiency of protection against babesia infection is at present a matter of speculation.

HUMPHREY: After recovery from babesia infection is the persistent immunity solid, or is there a brief parasitaemia following challeng If there is no detectable parasitaemia, does this imply that there is continual stimulation of immunity by persistent but undetected parasites?

ALLISON: Mice recovering from B. microti infections have persisten parasites detectable by transfer to non-immune recipients. Hence there is constant stimulation of the immune system, what used to be called premunition. But Callow has shown with bovine babesias that protective immunity can be obtained in the absence of persistent parasites.

HUMPHREY: How long does immunity induced by Corynebacterium parvum or BCG last? Does it correlate with the persistence of granulomata?

ALLISON: The immunity against babesias produced in mice after injection of killed C. parvum or live BCG takes about three weeks to become fully established. In the case of C. parvum, the immunity lasts for at least three months and thereafter declines. In the case of BCG it persists for at least six months, possibly because the living bacteria provide a persisting stimulus.

FORD: When making systematic measurements of natural killer activity is the choice of target cell critical? Would different target cells give discrepant results in some circumstances?

ALLISON: We use YAC tumour cells as targets so as to have results comparable to those of other laboratories in which this is the standard target cell strain. However, parallel differences in NK activity in mouse strains are manifested also with other cells as targets (for example, high activity in spleens of C57Bl mice and low activity in A mice). There are also comparable differences in the Cudkowicz assay for F1 reactivity against parental cells, but distinctions may of course emerge later.

MITCHELL: The diffusion chambers containing cells plus parasitized red cells may be a useful system with which to investigate the actions of purified T cells on parasitized cells. The prediction from Dr Sprent's presentation is that sensitized or virgin T cells will never be able to read the presence of antigen on a cell with such low H-2 expression as a mature erythrocyte.

ALLISON: That is one of the reasons why it seems unlikely that sensitized T cells can actually kill parasitized erythrocytes. The diffusion chamber experiments show that immune cells can exert their efforts at a distance, and the degeneration of parasites within circulating erythrocytes shows that lysis of infected erythrocytes or their opsonization for phagocytosis are not quantitatively important mechanisms of recovery from babesial infections. Parasite antigens may well be presented to immunocompetent cells or antigen-presenting cells (macrophages, veiled cells or dendritic cells).

HUMPHREY: Is there any evidence for an exo-erythrocytic stage in babesial infection?

ALLISON: Despite some claims, there is no solid evidence for a pre-erythrocytic or exo-erythrocytic stage in babesial infection.

COHEN: Have you looked in vivo for evidence of generalized modification of red cell permeability - for example, by introducing a second haemoprotozoan at the time of elimination of B. microti?

ALLISON: If mice recovering from B. microti are challenged with Plasmodium vinckei, which normally produces lethal infections, the plasmodia degenerate within circulating erythrocytes and the animals recover. So the cell permeability change appears to be non-specific.

MITCHELL: Can Dr Trager tell us whether a prominent feature of Plasmodium falciparum, cultivated under suboptimal nutritional circumstances, is a large number of degenerated intracellular parasite forms of the type described by Dr Allison in BCG/ B. microti infected mice?

TRAGER: Yes, in general, 'though we have mostly been interested in cultures maintained at optimal levels, with few degenerate forms.

The Role of the Spleen in Malaria Infections

David J. Wyler, Charles N. Oster and Thomas C. Quinn,
Laboratory of Parasitic Diseases, National Institutes of Health,
Bethesda, Maryland, USA.

INTRODUCTION

In the 4th century BC, more than 2000 years before Laveran discovered the etiological agent of malaria, the Greeks observed a relationship between the development of periodic fevers (malaria) and splenomegaly. The close association between the spleen and malaria may also have been the basis upon which Galen in 150 AD postulated that the function of the spleen was to extract black bile ("melancholy") from the blood. Perhaps the darkly pigmented spleens of patients who died of falciparum malaria stimulated these conclusions about the "organ full of mystery". Despite so auspicious a beginning, we find ourselves very much uncertain about either the physiological role of the spleen generally or about its mechanisms of host defence in malaria specifically.

The interest in the spleen in malaria has grown out of observations that no manipulation of the host's immune system so adversely affects the course of malaria as does splenectomy. Splenectomy not only alters the patterns of innate resistance of certain hosts to certain plasmodial species, but also results in the loss of protective immunity[1] in animals who have acquired this

---

[1] For the purposes of this discussion, we define protective immunity as the ability of the host to limit the magnitude of parasitaemia, to eliminate the parasite from circulation, or to be resistant to rechallenge.

as a result of infection or immunization. Since we know little about the critical parasiticidal mechanisms which provide for protective immunity in malaria, these observations on the effects of splenectomy have suggested the spleen as a potentially fertile ground for investigations in malaria immunology. However, the lack of an adequate technology for assessing splenic functions has retarded the rate of progress in this field. The unique architectural arrangement of spleen cells and the possibility that spleen structure might be important in malaria host defence has dictated a need for in vivo as well as in vitro experimental approaches. Furthermore, the intraerythrocytic location of the parasite suggests that the interaction of the spleen with infected erythrocytes might be determined by rheological, in addition to immunological, factors.

In this review, we will first explore some of the insights which have been gained simply through the analysis of the course of infection in animals following splenectomy. Next, we will review the state of knowledge concerning mechanisms of spleen function which may be important in host defence. Finally, we will consider certain aspects of immunoregulation in malaria as revealed by studies of spleen cells in vitro.

SPLENECTOMY AND THE COURSE OF MALARIA INFECTION

One way to learn about the functions of an organ in relation to a physiological process is to remove that organ and assess the effects of its ablation on that process. Such an approach opened the field of endocrinology and has been helpful in establishing the foundations of modern immunology. For us, it serves as an indispensable technique in investigating host defence in malaria.

An important feature of the biology of Plasmodia is the rather narrow range of hosts within which a species can develop. Some factors which determine host specificity have been identified which might be important determinants of innate resistance. These include certain nutritional factors, erythrocyte surface receptors, erythrocyte age, haemoglobin composition, and age of the host.(1) Less is known of the immunological factors which might contribute to such resistance. It has been observed, however, that certain experimental animals not normally susceptible to certain species of plasmodia can be rendered susceptible by splenectomy (e.g. Plasmodium gallinaceum from chickens into the rat). In other cases, infection with a parasite highly virulent to the natural host can result in brief, low grade infections in unnatural hosts (e.g. Plasmodium falciparum from man into Aotus trivirgatus). Splenectomy in the unnatural host may convert the infection to a malignant course that may mimic the one which occurs in the natural host. Without attempting to provide an exhaustive review of all such observations, we have summarized examples of increased susceptibility of splenectomized animals to certain plasmodial species to which they are not naturally susceptible (Table I).

TABLE I

THE EFFECT OF SPLENECTOMY ON THE SUSCEPTIBILITY OF ANIMALS TO INFECTION WITH VARIOUS PLASMODIUM SPECIES

PLASMODIUM SPECIES	NATURAL HOST	HOSTS SUSCEPTIBLE IF SPLENECTOMIZED
P. gallinaceum	Jungle fowl Domestic hen	Norwegian rat
P. vinckei	African thicket rat	Norwegian rat Hamster
P. chabaudi	African thicket rat	Norwegian rat
P. berghei	African thicket rat	Rhesus monkey Marmoset
P. cynomolgi	Macaque monkey	Owl monkey
P. falciparum	Man	Chimpanzee Owl monkey
P. vivax	Man	Chimpanzee Owl monkey Marmoset
P. malariae	Man Chimpanzee	Owl monkey
P. ovale	Man	Chimpanzee

Despite the many examples of altered innate resistance resulting from splenectomy, we have very little idea of how the spleen functions to protect these unnatural hosts. It does appear, however, that this protection is exclusively directed at the asexual intraerythrocytic forms of the parasite, since splenectomy appears to have no effect on susceptibility to sporozoite-induced infection (2,3) or on the development of exoerythrocytic stages. (4,5) There are also data suggesting the possibility that the presence of the spleen might affect the virulence of the parasite. If rhesus monkeys infected with Plasmodium cynomolgi are splenectomized, a strain of parasite subsequently appears which, upon inoculation into clean monkeys, is less virulent than the parent strain.(6) While these observations have not provided major insights into malaria immunology, they have provided a useful practical technique for isolation and study of new malaria species and strains in the laboratory. Perhaps in the future we will learn how the spleen of the unnatural hosts can be so precise in its recognition of certain parasite species, or how in susceptible hosts the parasites evade this recognition.

Investigations of the effects of splenectomy on malaria infection in animals which are susceptible in the intact state have revealed some interesting findings. In malaria models in which the host experiences a transient mild infection (limited parasitaemia and low mortality, as occurs in Plasmodium berghei infection of the adult rat) splenectomy prior to infection may result in severe, fatal infections (Figure 1, A).(7-10) The protective effect of the spleen in limiting the magnitude and duration of parasitaemia is not acquired by another reticuloendothelial organ, since even long delays between splenectomy and subsequent challenge result in severe infection.(6,11) In addition, animals with chronic malaria who can (as a result of infection) limit the magnitude of parasitaemia to low or subpatent levels experience rapid rises in parasitaemia and often overwhelming infections following splenectomy (Figure 1, C).(6,11) A dramatic example of this is the case of a patient who has chronic, asymptomatic Plasmodium malariae infection for 53 years.(12) Following an incidental splenectomy, he developed acute symptomatic malaria.

In an effort to more clearly define when during the course of malaria the spleen might exert important functions, we analyzed the effects of splenectomy on quartan malaria (due to Plasmodium inui) in rhesus monkeys.(11) Monkeys splenectomized prior to inoculation with Plasmodium inui had rapidly fatal infections associated with very high parasitaemia, whereas the intact monkeys developed benign infections characterized by low-grade parasitaemias and marked chronicity (greater than 13 years). In monkeys who were first infected and then splenectomized 2-3 months later, a very different pattern emerged. Although these animals rapidly developed the same levels of parasitaemia which killed monkeys splenectomized prior to infection, these monkeys tolerated their severe infections and splenectomy had no effect on mortality. Although unexplained, these observations raised the possibility that one function of the spleen

FIGURE 1

EFFECTS OF SPLENECTOMY ON THE COURSE OF MALARIA. PARASITAEMIA
IN INTACT ANIMALS (SOLID LINE) IS COMPARED WITH PARASITAEMIA
IN SPLENECTOMIZED ANIMALS (BROKEN LINES).
A, SPLENECTOMY PRIOR TO INOCULATION (INOC.); B, SPLENECTOMY
(SPLX) DURING INFECTION; C, SPLENECTOMY DURING LATENCY
LEADING TO ONE OF TWO POSSIBLE COURSES; D, SPLENECTOMY
DURING QUARTAN MALARIA (P. inui).

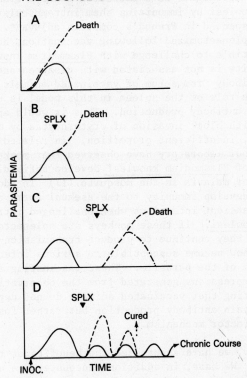

in acute malaria might be to provide for clinical tolerance to
the pathophysiological effects of high levels of parasitaemia.
Once the animal has experienced malaria, the spleen may be less
important in mediating tolerance.

Having considered the requirements for the spleen in defence during acute and chronic infection, let us now consider the role of the spleen in establishing immunity to reinfection. Animals which recover from an acute infection and clear the parasites from circulation are generally resistant to rechallenge with the homologous (and in some cases also heterologous) strain. In certain malarias, this may require repeated infections before immunity to reinfection is established. In some cases, this immunity is associated with the persistence of the parasite at subpatent levels. When such immune animals are splenectomized, they may lose their ability to resist reinfection.(13,14,15) Furthermore, the spleen plays an important role in the immunity stimulated by vaccination. Cohen and his colleagues have protected monkeys from infection with Plasmodium knowlesi by immunizing them intramuscularly with merozoites suspended in Freund's complete adjuvant. If these monkeys are splenectomized following vaccination, however, they become susceptible to challenge with Plasmodium knowlesi.(16) This loss of immunity is not associated with disappearance of anti-merozoite antibody from serum of vaccinated animals, indicating that the protective role of the spleen in this model is unlikely to be solely due to antibody production. Thus it would appear that antibodies which block invasion of erythrocytes by merozoites are, by themselves, insufficient protection. In related studies, Gwadz and Green in our laboratory have observed that monkeys immunized with gametes of Plasmodium knowlesi develop antibodies which block transmission of malaria in the mosquito.(17) In addition, these monkeys also develop immunity to the asexual forms and experience only mild, transient infections when challenged intravenously with Plasmodium knowlesi. If these monkeys are splenectomized following immunization, they continue to produce transmission-blocking antibody but now become susceptible to serious infection with the asexual stages of the parasite. These fascinating observations confirm the impressions generated from the observations of Cohen et al. suggesting that vaccinated animals do not depend upon the spleen to sustain antibody production but rather for some, as yet undefined, effector mechanism.

Until now, we have emphasized the benefits of the spleen in host defence. We have, in addition, encountered a unique and paradoxical situation which suggests that the spleen might also have deleterious effects on the host by promoting chronicity in certain infections. During the course of Plasmodium inui infection in rhesus monkeys intact animals left untreated had chronic, recrudescing infections which lasted for more than 13 years.(11) The rate at which this group of animals spontaneously cured themselves of infection (i.e. were persistently smear negative) was very low in the first few years after inoculation. In marked contrast, animals which were splenectomized and survived did not develop such chronic infections. Instead, they spontaneously cured themselves within the first year (Figure 1, D). Cure in some cases was confirmed by the failure of blood subinoculations from these animals to establish infection in non-immune recipients. The basis of these surprising findings is obscure at present. One possibility is that the spleen might be important in maintaining some form of immunological tolerance that results in recrudescences of parasitaemia. Such a hypothesis is stimulated by the evidence for spleen-dependent tolerance in other systems,(18,19) and if proven to be the explanation, would indicate a dichotomous immunological role of the spleen in this malaria model. Thus, the spleen might be beneficial at one phase of infection but have deleterious effects in addition to the beneficial effects, at another phase of infection. Since quartan malaria is unique in its marked chronicity, the deleterious effects of the spleen in relation to spontaneous cure may exist only in this and not other types of malaria.

In summary, the spleen exerts several influences on host defence against the intraerythrocytic, asexual stage of plasmodia: (1) it contributes to innate resistance of certain hosts to certain plasmodial species, (2) it acts to limit the magnitude of parasitaemia in acute infections, (3) it is required for restricting the level of parasitaemia in chronic infections, (4) it may be important for clinical tolerance early in infection, (5) it is required as an effector organ in mediating protection induced by vaccines but not for sustaining antibody production in response to these vaccines, and (6) it may play a role in establishing chronicity in certain malaria infections such as Plasmodium inui in the monkey. Having thus established the importance of the spleen in malaria, let us now explore possible mechanisms whereby this organ might exert its roles in these infections.

SPLENIC MECHANISMS OF HOST DEFENCE IN MALARIA

The spleen is capable of a variety of immunological and non-immunological functions which are likely to be of critical importance in malaria. The parasite is an obligate intraerythrocytic organism in the stage pathogenic for the host; the exoerythrocytic forms and gametocytes play no known role in the pathogenesis of the disease. As such, the inductive phase of immune response against the intraerythrocytic asexual form must occur primarily if not exclusively in the spleen, since it is unlikely that particulate parasite antigen finds its way to peripheral lymphoid tissue until late in infection, if at all.

Although the inductive phases of the immune response to malarial antigens have not been studied in any detail, we probably can generalize to this disease conclusions obtained in investigations of immune responses to heterologous erythrocytes. Rowley (20) observed that splenectomized rats injected intravenously with sheep erythrocytes produced much lower titres of haemolyzing antibody than did intact rats. When splenectomy was delayed until some time following sensitization, these animals produced normal amounts of antibody in response to a secondary challenge, since peripheral lymphoid organs could compensate for the spleen's removal. These studies indicated that the critical role of the spleen was during the primary antibody response to an intravenously-administered antigen, and that the spleen was not required thereafter for an anamnestic response. Since no such requirement for the spleen has been observed when the antigen is administered subcutaneously or intramuscularly, it is clear that the route of antigen administration is an important determinant in the requirement for the spleen in the primary immune response.

Although antibodies of the IgG class might be the most important in protective immunity in malaria (at least in rodents) (21) it is of interest that relapsing infections in humans are associated with repeated boosts in IgM (22) rather than stimulating an anamnestic response. A possible explanation for this observation is that each relapse (or recrudescence) represents a population of parasites which contain unique antigenic determinants (23) that stimulate primary immune responses. This would suggest that splenectomy, even in the chronic stages of infection, might continue to be important in the induction of immune responses, and why peripheral lymphoid organs might not be able to compensate for splenic ablation.

It is uncertain at present what the role of antibody is in mediating host defence in malaria. Investigators have provided evidence that serum and even immunoglobulin transfer from immune animals can impart protection to the recipients.(21,24) Furthermore, ablation of B-cell function by the treatment with anti-μ antibody resulted in more severe Plasmodium yoelii infection in mice,(25) also suggesting an important role of B cells (and, presumably, antibody) in host defence. In vitro studies (26) have suggested that antibody might function to block the invasion of erythrocytes, and yet Miller et al. (27) were unable to find a correlation between the immune status of monkeys in vivo and the presence of such blocking antibodies as tested in vitro. More recently, it has been shown that the presence of circulating antimerozoite antibodies per se was insufficient to protect vaccinated monkeys against Plasmodium knowlesi; an intact spleen

was an additional requirement.(16)   In addition, only intact young
rats could be protected from lethal infections of Plasmodium berghei
by the administration of immune serum; immunity could not be
transferred to splenectomized rats.(28)

Another potential role of antibody in malaria is as an
opsonizing or cytophilic antibody which could promote phagocytosis
of infected erythrocytes. The histopathological evidence for the
occurrence of marked phagocytosis of parasite material, first
recognized nearly 100 years ago by Golgi, led Taliaferro and
Mulligan (29) to postulate that opsonization and phagocytosis was
the major host defence mechanism in malaria. Brown et al. (30)
tested this by studying macrophage phagocytosis of Plasmodium
knowlesi-infected erythrocytes in vitro. They observed that in the
presence of immune serum schizonts were ingested by macrophages.
It is uncertain however, whether erythrocytes containing younger
stages of the parasites can also be opsonized. Possibly parasite
antigens are only expressed on the surface of erythrocytes
containing mature parasites. On the other hand, the exposure of
parasite antigens on the surface of erythrocytes containing younger
forms of the parasite might theoretically occur if these cells were
first minimally damaged. Precedent for minimal damage of
erythrocytes can be found in the case of certain haemolytic
anaemias in which the surface of abnormal erythrocytes is altered
by the spleen in a process called "conditioning". The membrane
defects resulting from red cell conditioning may culminate in the
phagocytosis of these altered cells.(31)   In malaria, such
conditioning might result in exposure of parasite antigens and
thereby provide for the subsequent opsonization of the infected
red cell.

One way in which parasitized red cells can interact with the
spleen in a non-immunologically dependent mechanism is through a
process known as "pitting". This process serves as an example of
the way in which the spleen can interact with parasitized red
cells on the basis of altered rheological properties of these cells.
The process, first described to explain the removal of other
intraerythrocytic particles such as Heinz bodies, iron particles
and Howell-Jolly bodies,(32) has also been observed in malaria.(33,
34)   In malaria, infected erythrocytes course through the splenic
cords into the sinuses, passing through narrow fenestrations.
Unlike normal erythrocytes, infected erythrocytes are not highly
deformable (35) and as they pass through these fenestrations the
portion of the cell containing the parasite becomes trapped
(Figure 2). The portion of the red cell which has successfully
passed through the fenestration separates from the part containing
the parasite and its membrane reseals, creating a spherocyte. The

192

FIGURE 2

PITTING IN SPLENIC CORDS. (SEE TEXT FOR EXPLANATION)
NRBC, NORMAL ERYTHROCYTE. IRBC, INFECTED ERYTHROCYTE.
S, SPHEROCYTE. P, PARASITE

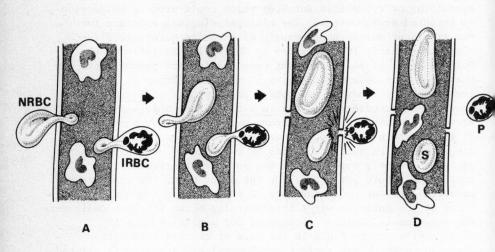

parasitized portion is presumably quickly ingested by resident
macrophages while the spherocyte is cleared from circulation at a
later time. It is unlikely that pitting contributes in an
important way to host defence in malaria since this process lacks
a recognition system which would provide for the species-specificity
of malaria immunity. Furthermore, since this process occurs in
both immune and non-immune animals, it is unclear how immunity
could depend upon such a mechanism. Possibly alterations in
microcirculatory pathways or in the effective size of the cordal
fenestrations could modulate the degree to which pitting occurs, but
its importance in malaria is yet to be established.

In an effort to assess how immune and rheologic mechanisms
interact in the spleen, we have begun to examine factors which
regulate the removal of parasitized erythrocytes from circulation.
We are studying the kinetics of disappearance of intravenously
administered ^{51}Cr-labelled viable <u>Plasmodium berghei</u>-infected

erythrocytes from the circulation of inbred rats and are also determining the rate and magnitude of organ localization of these infected cells (Quinn and Wyler, unpublished data). We observed that in uninfected control rats, approximately 10% of the parasitized erythrocytes were removed from circulation and localized in the spleen within the first hour following inoculation. In rats rendered immune to reinfection by virtue of prior infection with Plasmodium berghei, 30% of the inoculum was cleared by the spleen in the first hour. These differences were directly related to the greater size of the spleen in immune animals, and did not represent a greater efficiency of uptake of infected red cells by immune spleens. After the first hour following inoculation, clearance rates of tagged cells were the same in immune and non-immune rats, although parasitaemia fell in the immune and rose in the non-immune rats. This indicated that protective immunity in immune rats was not simply due to accelerated intravascular clearance of parasitized erythrocytes. Clearance rates were identical in splenectomized immune and splenectomized control rats, indicating that clearance differences between these groups in the intact state were spleen-dependent. We interpreted these data as evidence against antigen-specific opsonization in vivo, since the occurrence of such a process should have revealed differences between the animals which were not simply related to spleen size. To test this interpretation, we pre-incubated infected erythrocytes in immune serum prior to tagging with ^{51}Cr and injecting into control rats. This procedure failed to alter the pattern of clearance of the tagged cells. Next, we compared the clearance of uninfected rat red cells coated with subagglutinating concentrations of rabbit anti-rat erythrocyte antibody in immune and non-immune rats. We observed that although clearance of these cells were much more rapid than parasitized erythrocytes, there was no significant difference between the clearance patterns of these cells in immune and non-immune rats. This indicated that clearance of parasitized erythrocytes in immune animals is not analogous to clearance of antibody-coated red cells and therefore was unlikely to be antibody-depdendent. Rather, it seemed that rheologic determinants may be important in splenic clearance of infected erythrocytes. The clearance patterns of parasitized red cells is similar to the clearance of Heinz-body containing erythrocytes in that both are characterized by rapid component in the first hour followed by slower subsequent components, both have a major dependence on the spleen for their clearance, and, the clearance of both in immune and control rats is proportional to the spleen size.

When we examined clearance at various times in the course of infection, two additional findings appeared. First of all, it became apparent that on-going infection imparts some degree of reticuloendothelial blockade. Second, during the period of "crisis" when the infection begins to resolve, there is a sudden and dramatic increase in clearance of parasites to the spleen. We are at present unable to explain these observations, but we are pursuing studies which we hope will elucidate the mechanisms underlying these sudden changes in clearance.

Our evidence in the rat model that non-immunological mechanisms might play an important role in host defence raised the possibility that the architectural configuration of the spleen - and not merely the cell populations located in this organ - might be an important determinant. To test this hypothesis, we splenectomized mice and immediately reconstituted them with either intraperitoneal inoculations of autologous spleen cell suspensions or subcutaneous implanatation of autologous spleen chunks.(Oster and Wyler, unpublished observations) After several weeks, we challenged these mice with a non-lethal strain of Plasmodium yoelii and compared the course of infection in these mice with that in intact and splenectomized, non-reconstituted mice (Figure 3). In preliminary studies we observed that reconstitution protected the mice from the lethal effects of splenectomy. The mice reconstituted with spleen chunks regenerated small splenules with normal splenic architecture. Mice reconstituted with cell suspensions, however, developed no identifiable splenic tissue and we presume the cells populated the bone marrow and other sites. This suggests that at least for Plasmodium yoelii in the mouse, splenic architecture is probably not critically important for resolution of the acute infection. Rather, some yet-to-be-identified spleen cell population is the critical element. In addition to attempting to identify that cell population, we are determining whether a similar conclusion can be made for Plasmodium berghei in the rat.

In summary, it appears that the spleen must play an important role in immune induction in malaria, and can also exert non-immunological parasiticidal effects. In vivo evidence for antibody-mediated intravascular clearance of parasitized erythrocytes is lacking in the rodent malaria model. Whether an interaction between rheologically determined splenic "conditioning" mechanisms and subsequent opsonization might be required for protective immunity must be determined. Preliminary data suggest that, in the mouse, a spleen cell population is present which is critical for the resolution of acute, non-lethal malaria, and that in this model there may be no apparent requirement for an architecturally intact spleen. Let us now focus on certain aspects of the spleen in malaria as viewed at the cellular level.

FIGURE 3

THE EFFECTS OF SPLENIC RECONSTITUTION ON THE COURSE OF P. YOELII
MALARIA IN MICE. A, INFECTION IN INTACT MICE; B, INFECTION IN
SPLENECTOMIZED MICE; C, INFECTION IN SPLENECTOMIZED MICE
RECONSTITUTED WITH AUTOLOGOUS SPLEEN CHUNKS; D, INFECTION IN
SPLENECTOMIZED MICE RECONSTITUTED WITH
AUTOLOGOUS SPLEEN CELL SUSPENSIONS.

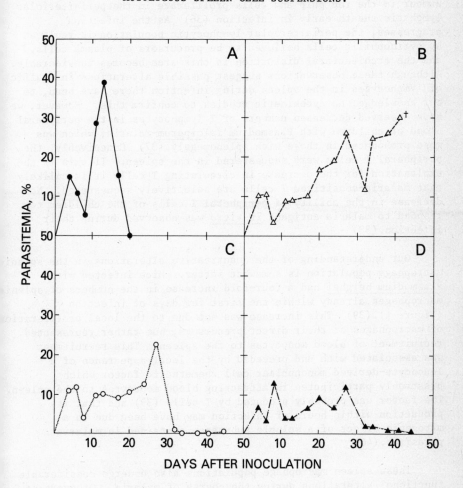

ALTERATIONS IN SPLEEN CELLS DURING MALARIA

The major histopathological alterations which occur during the course of malaria are primarily in the spleen and, to a lesser extent, in the liver. Parasitized erythrocytes and uninfected red cells are sequestered in the splenic red pulp where they and their products are ingested by tissue macrophages. In addition, rodents infected with malaria show histological evidence of stimulated erythropoiesis occurring in the spleen. Macrophages increase in number in the red pulp and cells proliferate in the periarteriolar lymphatic sheath early in infection.(36) As the infection progresses, the periarteriolar lymphocytic population is replaced by pyrinophilic cells believed to be precursors of plasma cells, and the architectural distortion in this area becomes considerable. Although these observations suggest possible alterations in traffic of lymphocytes in the spleen during infection there have been, to our knowledge, no cytokinetic studies to confirm this. However, we have observed decreased numbers of T lymphocytes in the peripheral blood of children with Plasmodium falciparum-malaria, which was most pronounced in those with splenomegaly.(37) Conceivably, the peripheral T cells were sequestered in the spleen. If this is the explanation for the decrease in circulating T cells it is unlikely that malaria-sensitized T cells are selectively sequestered. No decrease in the ability of peripheral T cells of the children to respond to malaria antigens in vitro was observed during their infection.(38)

Our understanding of the quantitative alterations in the splenic macrophage population is somewhat better. Mice infected with Plasmodium berghei had a four-fold increase in the numbers of splenic macrophages already within the first few days of infection (Figure 4).(39) This increase was not due to the local proliferation of macrophages or their direct precursors, but rather represented recruitment of blood monocytes to the spleen. This recruitment was associated with and preceded by the local appearance of a leucocyte-derived mononuclear cell chemotactic factor which presumably participated in attracting blood monocytes to the spleen. The factor was probably produced by T cells (39) and that its production within hours of infection may have been due to a mitogenic effect of a soluble substance contained in malaria parasites.(40)

These spleen macrophage populations also undergo considerable functional alterations during the course of malaria. Of greatest interest, and best studied, are the role these cells play in immunoregulation during malaria. Loose et al. (41) first suggested that the immunosuppression which occurs in malaria is due to a

FIGURE 4

THE RELATIONSHIP BETWEEN APPEARANCE OF MONONUCLEAR CELL
CHEMOTACTIC FACTOR IN THE SPLEEN AND INCREASE IN THE SPLEEN
MACROPHAGE POPULATION IN MICE INFECTED WITH P. BERGHEI.
ABCISSA SHOWS TIME AFTER INTRAVENOUS INOCULATION OF PARASITES.[1]

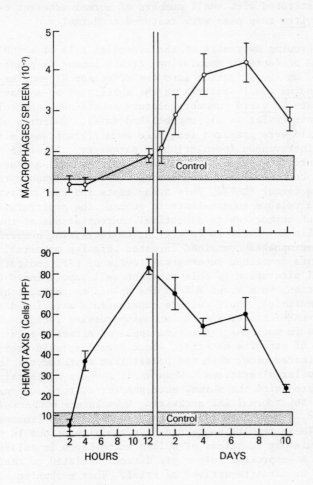

---

[1] (39: Reproduced by permission of the Journal of Immunology).

defect in macrophage function, specifically the inability of these cells to process and focus antigen. Warren and Weidanz (42) subsequently studied this effect in vitro and demonstrated that spleen cells of malaria-infected mice were deficient in mounting a primary in vitro antibody response to horse red cells. When the malarious spleen cells were depleted of adherent cells (macrophages) and reconstituted with small numbers of normal adherent cells, these in vitro responses were restored to normal.

The growing awareness of the potential role of soluble macrophage products in modulating certain immune responses stimulated an investigation into the effects of Plasmodium berghei and Plasmodium yoelii malaria on the ability of mouse macrophages to elaborate in vitro immunoregulatory substances into culture supernatants (Wyler et al. unpublished data). Cultures of adherent spleen cells were prepared from mice on different days after a standard intravenous inoculation of parasites. After 48 hours of cell culture incubation, supernatants were removed and assayed for the presence of the immunoaugmenting substance, lymphocyte activating factor (LAF). This assay tests for the ability of macrophage culture supernatants to augment the proliferative response of thymocytes to suboptimal concentrations of the mitogen Concanavalin A. Early in infection, as increasing numbers of splenic macrophages contained ingested parasite material, culture supernatants contained supernormal levels of LAF, indicating a functional alteration in this adherent cell population. Later in the infection, when about 80% of the spleen macrophages had ingested parasite material, their culture supernatants were found to contain subnormal LAF activity and an immunosuppressive factor. This modulation in macrophage function appeared directly related to the ingestion of parasite material, since normal splenic macrophages from uninfected mice, when fed parasitized erythrocytes in vitro, showed similar alterations. When small numbers of parasites were co-cultivated with the normal macrophages, culture supernatants contained supernormal LAF activity. When larger numbers of parasites were added, the culture supernatants contained the immunosuppressive factor. These results suggested that the alterations in the immunoregulatory functions of spleen macrophages in malaria – from a helper to a suppressor role – are directly related to their ingestion of parasite-derived material. This mechanism might explain how spleen macrophages mediate the immunosuppression of malaria. It appears that this suppressor function is compartmentalized primarily in the spleen since one can readily demonstrate suppression of spleen lymphocyte responses (42,43) but not suppression of peripheral blood lymphocytes responses (38) during malaria infection. The compartmentalization of immune suppression is perhaps best explained by the greatest opportunity for parasite-macrophage interaction occurring in the spleen and the resulting conversion of these macrophages to an immunosuppressive role.

Very little is known about functions of spleen cell populations in mediating protective immunity. We have already referred to the present state of knowledge of phagocytosis in malaria and to our uncertainties about the overall importance of phagocytosis as a host defence mechanism in this disease. Coleman and his colleagues have presented evidence that a spleen cell population might be capable of mediating a cytotoxic effect on infected erythrocytes. (44) They observed that the co-cultivation of spleen cell suspensions from malaria-immune but not from control mice caused release of ^{51}Cr from infected erythrocytes in vitro. These observations must be verified and extended to provide some idea of the relative importance of cytotoxicity in host defence before we can conclude that it is a major determinant in protective immunity.

Clark et al. (45) have observed that intravenous administration of BCG protected mice from Plasmodium vinckei infection and that in this setting the intracellular circulating parasites appeared morphologically abnormal. They postulated that the injected BCG stimulated a cell population (possibly macrophages) to secrete a soluble substance in some way caused intracellular death or failure of development of the parasite. The role which the spleen plays in mediating this effect is uncertain, although we would anticipate that the spleen may be necessary for this effect. If naturally-acquired immune mechanisms function in a similar manner, then the species specificity which characterizes these mechanisms must be explained. Perhaps specificity is imparted by events which trigger this mechanism or by additional factors such as specific antibody interacting with the non-specific component. The hypothesis proposed by the investigators needs to be related to naturally acquired immunity.

CONCLUSIONS

It is clear from this discussion that no integrated schema can be devised at present to explain the role of the spleen in malaria. The various ways in which splenectomy can alter host defence in malaria are being recognized. The spleen possesses a highly specific recognition system, since it acts to prevent infection with certain species of malaria in intact hosts, while at the same time not preventing infection with other species of malaria. It plays

an important role in the induction and expression of acquired
protective immunity and may also play a role in clinical tolerance
to certain pathophysiological processes which occur in malaria.
There is also an absolute need for the spleen in the expression of
immunity acquired by vaccination.

The complex mechanisms which mediate these important effects
are poorly understood. While the induction of primary antibody
response to an intravascular particulate antigen is clearly spleen-
dependent, there is evidence that the spleen is not required for
persistent antibody production in malaria. However, it is
uncertain whether the production of specific IgM directed at variant
populations of parasites emerging during relapses or recrudescences
is spleen-dependent. Too little is known about the importance of
opsonization as a parasiticidal mechanism to be certain of the
spleen's role in this process. In rodent malaria (Plasmodium
berghei in the rat) antibody-mediated clearance and opsonization of
intact infected erythrocytes are probably not important defence
mechanisms. Whether this is true in other malaria infections must
be determined. How other mechanisms might contribute to host
defence, such as interactions of the spleen with infected
erythrocytes on the basis of their altered rheologic properties,
must be more extensively explored. Finally, it is clear that
immunological perturbations occur in association with malaria
infection and that functional alterations in spleen-cell populations
may be responsible.

How does this all relate to the global problem of malaria
control? Prospects for vaccination against malaria may depend upon
the gaining of an understanding of host defence mechanisms and
immunoregulation in malaria. Vaccines at present available which
successfully prevent infectious diseases are, by and large, directed
against diseases in which protective immunity also develops as a
consequence of natural infection. In contrast, malaria in man is
characterized by sluggish and incomplete development of immunity.
Whether this is due to the immunosuppression associated with malaria
has yet to be determined. In any event, employing the classical
empiric approach to vaccination which has been so successful in
other diseases may not ultimately prove successful in human malaria.
It may be necessary to first learn how the immune host protects
itself in malaria and how the parasite evades these mechanisms,
before we can hope to successfully prevent malaria through
vaccination. In view of the substantial evidence for a central role
of the spleen in host defence and immunoregulation in malaria, it
seems advisable to intensify our investigations of this "organ full
of mystery".

REFERENCES

1. MILLER, L.H. and CARTER, R. Innate resistance in malaria. Experimental Parasitology, 40: 132 (1976).

2. BAFORT, J.M. Etude du cycle biologique du Plasmodium vinckei. IV. Virulence et immunité. Annales de Sociétés Belges de Médecine Tropicale, 49: 610 (1969).

3. BAFORT, J.M. Role of the spleen in Plasmodium vinckei with particular reference to splenectomy. Annales de Sociétes Belges de Médecine Tropicale, 51: 169 (1971).

4. GARNHAM, P.C.C. and BRAY, R.S. The influence of immunity upon the stages of mammalian malaria parasites. Revista Brasiliera de Malariologia e Doenças Tropicais, 8: 151 (1956).

5. BRAY, R.S. Studies on malaria in chimpanzees. IV. Plasmodium ovale. American Journal of Tropical Medicine and Hygiene, 6: 638 (1957).

6. GARNHAM, P.C.C. The role of the spleen in protozoal infections with special reference to splenectomy. Acta Tropica (Basel), 27: 1 (1970).

7. ZUCKERMAN, A. and YOELI, M. Age and sex as factors influencing Plasmodium berghei infections in intact and splenectomized rats. Journal of Infectious Diseases, 94: 225 (1954).

8. PRAKESH, S. Studies on Plasmodium berghei. Effects of splenectomy on the course of blood-induced infections in rats. Indian Journal of Malariology, 15: 107 (1961).

9. CANTRELL, W. and MOSS, W.G. Partial hepatectomy and Plasmodium berghei in rats. Journal of Infectious Diseases, 113: 67 (1963).

10. SPIRA, D. and ZUCKERMAN, A. Blood loss and replacement in plasmodial infections. VI. Plasmodium berghei in splenectomized rats. Journal of Infectious Diseases, 115: 337 (1965).

11. WYLER, D.J. ET AL. Spleen function in quartan malaria (due to Plasmodium inui): Evidence for both protective and suppressive roles in host defence. Journal of Infectious Diseases, 135: 86 (1977).

12. GUAZZI, M. and GRAZI, S. Considerazioni su un caso do malaria quartana recidivante dopo 53 ani di latenza. Rivista di Malariologia, 42: 55 (1963).

13. CORRADETTI, A. Particolari fenomeni immunitari nell infezione de Plasmodium berghei. Rivista di Parsitologia, 11: 201 (1950).

14. COX, F.E.G. Acquired immunity to Plasmodium vinckei. Progress in Protozoology, 91: 167 (1965).

15. ZUCKERMAN, A. Residual immunity following radical cure of Plasmodium berghei in intact and splenectomized voles (Microtus guentheri). Journal of Infectious Diseases, 92: 205 (1953).

16. BUTCHER, G.A. ET AL. Antibody mediated mechanisms of immunity to malaria induced by vaccination with Plasmodium knowlesi merozoites. Immunology, 34: 77 (1978).

17. GWADZ, R.W. and GREEN, I. Malaria immunization in rhesus monkeys: A vaccine effective against both the sexual and asexual stages of Plasmodium knowlesi. Journal of Experimental Medicine (in press).

18. MACKANESS, G.B. ET AL. Feedback inhibition of specifically sensitized lymphocytes. Journal of Experimental Medicine, 139: 543 (1974).

19. SY, M.-S. ET AL. A splenic requirement for the generation of suppressor T cells. Journal of Immunology, 119: 2095 (1977).

20. ROWLEY, D.A. The effect of splenectomy on the formation of circulating antibody in the adult male albino rat. Journal of Immunology, 64: 289 (1950).

21. DIGGS, C.L. and OSLER, A.G. Humoral immunity in rodent malaria. II. Inhibition of parasitemia by serum antibody. Journal of Immunology, 102: 298 (1969).

22. TOBIE, J.E. ET AL. Serum immunoglobulin levels in human malaria and their relationship to antibody production. Journal of Immunology, 97: 498 (1966).

23. VOLLER, A. and ROSSAN, R.N. Immunological studies with simian malarias. I. Antigenic variants in Plasmodium cynomolgi bastianelli. Transactions of the Royal Society of Tropical Medicine and Hygiene, 63: 46 (1969).

24. COHEN, S. ET AL. Gamma-globulin and acquired immunity to human malaria. Nature, 192: 733 (1961).

25. WEINBAUM, F.I. ET AL. Immunity to Plasmodium berghei yeolii in mice I. The course of infection in T cell and B cell deficient mice. Journal of Immunology, 117: 1999 (1976).

26. COHEN, S. ET AL. Action of malarial antibody in vitro. Nature, 223, 368 (1969).

27. MILLER, L.H. ET AL. Plasmodium knowlesi: Functional immunity and antimerozoite antibodies in Rhesus monkeys after repeated infection. Experimental Parasitology, 41: 105 (1977).

28. BROWN, I.N. and PHILLIPS, R.S. Immunity to Plasmodium berghei in rats: Passive serum transfer and role of the spleen. Infection and Immunity, 10: 1213 (1974).

29. TALIAFERRO, W.H. and MULLIGAN, H.W. The histopathology of malaria with special reference to the function and origin of the macrophages in defence. Indian Medical Research Memoirs, 29: 1 (1937).

30. BROWN, K.N. ET AL. Immunity to malaria. II. Serological responses of monkeys sensitized by drug suppressed infection or by dead parasitized cells in Freund's complete adjuvant. Experimental Parasitology, 28: 318 (1970).

31. COOPER, R.A. ET AL. Role of the spleen in membrane conditioning and hemolysis of spur cells in liver disease. New England Journal of Medicine, 290: 1279 (1974).

32. CROSBY, W.H. Normal functions of the spleen relative to red blood cells: a review. Blood, 14: 399 (1959).

33. CONRAD, M.E. and DENNIS, L.H. Splenic function in experimental malaria. American Journal of Tropical Medicine and Hygiene, 17: 170 (1968).

34. SCHNITZER, B. An ultrastructural study of the red pulp of the spleen. Blood, 41: 207 (1973).

35. MILLER, L.H. ET AL. Alteration in the rheologic properties of Plasmodium knowlesi-infected red cells. A possible mechanism for capillary obstruction. Journal of Clinical Investigation, 50: 1451 (1971).

36. MORAN, C.J. ET AL. Immunological significance of histological changes in the spleen and liver in mouse malaria. Clinical and Experimental Immunology, 13: 467 (1973).

37. WYLER, D.J. Peripheral lymphocyte subpopulations in human falciparum malaria. Clinical and Experimental Immunology, 23: 471 (1976).

38. WYLER, D.J. and BROWN, J. Malaria antigen-specific T cell responsiveness during infection with Plasmodium falciparum. Clinical and Experimental Immunology, 29: 401 (1977).

39. WYLER, D.J. and GALLIN, J.I. Spleen-derived mononuclear cell chemotactic factor in malaria infections: A possible mechanism for splenic macrophage accumulation. Journal of Immunology, 118: 478 (1977).

40. WYLER, D.J. and OPPENHEIM, J.J. Lymphocyte transformation in human Plasmodium falciparum malaria. Journal of Immunology, 113: 449 (1974).

204

41. LOOSE, L.D. ET AL. Malarial immunosuppression – a macrophage mediated defect. Proceedings of the Helminthological Society of Washington, 39 (Special issue): 484 (1972).

42. WARREN, H.S. and WEIDANZ, W.P. Malarial immunodepression in vitro: adherent spleen cells are functionally defective as accessory cells in the response to horse erythrocytes. European Journal of Immunology, 6: 816 (1976).

43. SPIRA, D.T. ET AL. The reactivity of spleen cells from malarious rats to non-specific mitogens. Clinical and Experimental Immunology, 24: 139 (1976).

44. COLEMAN, R.M. ET AL. Splenic mediated erythrocyte cytotoxicity in malaria. Immunology, 29: 49 (1975).

45. CLARK, I.A. ET AL. Protection of mice against Babesia and Plasmodium with BCG. Nature, 259: 309 (1976).

Preliminary Investigations on the Role of the Spleen in Immunity to *Plasmodium Knowlesi* Malaria

J. Langhorne, G.A. Butcher, G.H. Mitchell and S. Cohen,
Department of Chemical Pathology, Guy's Hospital
Medical School, London, England

INTRODUCTION

Rhesus monkeys vaccinated with merozoites in Freund's complete adjuvant (FCA) are protected against otherwise lethal challenge with distinct variants and strains of Plasmodium knowlesi.(1-3) The mechanisms of immunity induced by this vaccine are not completely understood. Correlation between clinical immune status and merozoite-inhibitory antibody is not invariable. Other cell-mediated specific and non-specific mechanisms appear to be induced and the cells involved may be splenic in origin.(4)

The importane of the spleen in malaria has long been recognized (5) but its precise role is not known. This paper describes the efficacy in the rhesus monkey of merozoite vaccination in the absence of the spleen, and effects of splenectomy on the clinical immunity of immunized rhesus monkeys and marmosets which were previously resistant to repeated challenge infections with different variants of P. knowlesi.

Although in vivo experiments suggest the importance of cell mediated effector mechanisms in the immunity to malaria,(6) there are no satisfactory in vitro assays for the detection and definition of such processes. The results of preliminary experiments on the in vitro growth of P. knowlesi in the presence of immune spleen cells from rhesus and kra monkeys are presented in this paper.

MATERIALS AND METHODS

Imported rhesus monkeys (<u>Macaca mulatta</u>), kra monkeys (<u>Macacca fasicularis</u>), and common marmosets (<u>Callithrix jacchus</u>) were caged in a room artifically illuminated between 5 a.m. and 5 p.m. The strains and variants of P. knowlesi parasites, isolation of merozoites, vaccination, challenge and detection of parasites have been described previously.(1-3,7)

INHIBITORY ANTIBODY

The antibody inhibiting invasion of erythrocytes by merozoites was assayed <u>in vitro</u> using cultures of defined parasite variants. (8,9)

Because of the small volumes of sera available from marmosets, inhibitory antibody was determined in a semi-micro modification of the previously described test. Briefly, 200µl of a 10% erythrocyte suspension, containing 1% parasitized cells (schizont stage) in medium RPMI 1640 (Gibco Biocult, Renfrewshire, Scotland) supplemented with 20mM Tes, 50 i.u. neomycin and 10% normal serum, were incubated in flat bottomed micro-titre plates (Flow labs., Irving, Scotland) for 18 hours in the presence of 20µl of test serum. The percentage inhibition of reinvasion was determined by parasite counts on the blood films stained with Giemsa.

PREPARATION OF LEUCOCYTES

Spleen cells

A single cell suspension was prepared by gently mincing the spleen in sterile medium RPMI and passing it through a fine mesh sieve. After one wash in a large volume of medium, the cells were incubated at room temperature for 5 minutes in $0.75\% NH_4Cl$ (^{w}/v) in 0.01M Tris buffer pH 7.4. After incubation the cells were washed a further two times, resuspended at 5×10^6 cells/ml in medium containing 10% normal serum and incubated overnight at $37^\circ C$, in an atmosphere of 5% CO_2. Before culture with P. knowlesi parasites, they were washed twice in medium. Viability of the cells as assessed by a Trypan Blue dye exclusion test was generally greater than 85%. The identification of the spleen cells was carried out as described below.

Peripheral blood mononuclear cells

Peripheral blood mononuclear cells were prepared from whole heparinised monkey blood by centrifugation over a Ficoll-Triosil mixture as described by Böyum.(10) Viability of the cells after washing three times in medium was greater than 98%.

Identification of leucocytes

Polymorphonuclear and mononuclear cells were assessed by differential counts on cell smears stained with Giemsa. An E-rosette test with sheep erythrocytes was used as a T-cell marker. (11) B cells were identified by surface immunoglobulins using a sheep anti-human Ig-FITC conjugate (Burroughs Wellcome, Beckenham, Kent). Monocytes/macrophages were identified by morphology and by a non-specific Esterase stain.(12)

Fractionation of spleen cells

A B-cell enriched fraction was obtained by eliminating E-rosettes on a Ficoll-Triosil gradient as described previously.(13) The E-rosette enriched fraction was recovered from the Ficoll-Triosil centrifugation by lysis of sheep erythrocytes in $0.75\%NH_4Cl/0.01M$ Tris buffer. The recovered cells were washed three times prior to use, and cells in each fraction were identified as described above.

In vitro culture of P. knowlesi parasites with immune and normal leucocyte preparations

Washed parasitized erythrocytes (ring stage parasites) from an infected donor monkey were mixed with washed normal monkey erythrocytes to give an approximate parasitaemia of 1%. This parasitized blood was diluted in medium RPMI 1640 supplemented with 20mM Tes, 0.4% sodium bicarbonate, 50 i.u. neomycin, and 20% normal monkey serum, to give a final concentration of 5×10^8 erythrocytes/ ml (approximately 5×10^6 parasites/ml). Leucocyte suspensions from immune and normal monkeys diluted in medium containing 20% normal serum to give effector to target ratios of 5:1, 2.5:1 and 1:1 (leucocyte:parasite). One hundred µl of the parasitized erythrocyte suspension was added to the different concentrations of leucocytes in flat bottomed micro-titre trays, and 100µl of the medium containing 20% serum was added to each well. The trays were incubated for 24-30 hours at $37^{\circ}C$ in a 5% CO_2 atmosphere. The parasitaemias after incubation were assessed by counting parasites on thin blood films stained with Giemsa. Each experiment was performed in duplicate or triplicate, and the results expressed as the mean parasitaema $\pm$ standard error.

RESULTS

Vaccination and challenge of splenectomized rhesus monkeys

Two rhesus monkeys were splenectomized three days prior to vaccination with W1 merozoites of P. knowlesi in FCA. These animals were subsequently challenged with the homologous variant of the parasite (Table I). Parasites were first observed in the peripheral blood 6 days after challenge. Maximum parasitaemias of 23% and 52% were recorded, and both animals died 14 days after challenge. Prepatent periods were not substantially prolonged compared with those obtained with control animals challenged with the same number of P. knowlesi parasites. However, the course of the patent parasitaemias in the two splenectomized vaccinated monkeys differed from that observed in the controls, in that the initial parasite multiplication was not logarithmic, and the day of death was delayed (Fig. 1). Antibody inhibiting reinvasion of erythrocytes by W1 merozoites in vitro was induced by vaccination of these splenectomized animals and reached levels of 29% and 49% prior to challenge (Table I).

Splenectomy and challenge of merozoite immunized rhesus monkeys

The results of challenge infections in six rhesus monkeys vaccinated with W1 merozoites in FCA are summarized in Table II. Monkeys vaccinated in this manner eliminate challenge after brief parasitaemias, in contrast to challenge of normal rhesus monkeys which results in a fatal malaria within nine days in approximately 90% of cases.(1,3)

These monkeys, resistant to repeated challenge with homologous (W1) or heterologous (W3) variants of P. knowlesi were splenectomized (Table III). Five monkeys were challenged one day prior to or on the day of splenectomy. Three monkeys survived this early challenge infection and developed only low-grade patent parasitaemias of short duration (Table III - MZ22, MZ42 & BW283). Two monkeys died of a fulminating malaria infection within 11 days of challenge. The prepatent periods were not extended, but in one animal (MZ64) the parasite multiplication was not logarithmic. Both of these monkeys had produced levels of inhibitory antibody against W1 or W3 variants of P. knowlesi before splenectomy which were comparable to those in the three surviving animals (Table III). Four monkeys were given a late challenge with W1 or W3 variants four to six weeks after splenectomy. Two animals survived challenge infection developing negligible or low-grade parasitaemias of short duration (Table III - MZ42 & BW283). MZ42 was challenged again 56 days

TABLE I

VACCINATION OF PREVIOUSLY SPLENECTOMIZED RHESUS MONKEYS WITH P. KNOWLESI (W1) MEROZOITES (Mz) IN FACA: RESULTS OF CHALLENGE INFECTION

Monkey	Vaccination			Challenge		Parasitaemias				Inhibitory antibody titres	
	No. Mz (variant)	Adjuvant[2]	days[2]	Variant	day[2]	Prepatent period (days)	Duration of patency (days)	Maximum parasitaemia (per 10^4 rbc)	Day of death (post challenge)	Anti W1 %	W1 (day)[2]
MZ65[1]	2×10^9 (W1)	FCA	3,33	W1	66	4	10	5200	14	0 49	(3) (47)
MZ66[1]	2×10^9 (W1)	FCA	3,33	W1	66	4	10	2300	14	0 29	(3) (47)
G551	–			W1	0	4	5	5000	9		
G552[1]	–			W1	8	3	5	5100	8		

(1) MZ65, MZ66 and G552 were splenectomized on day 0

(2) Indicates days after splenectomy

FIGURE 1

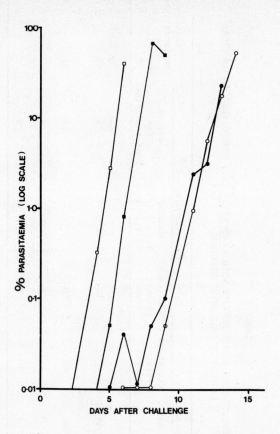

Parasitaemia following intravenous challenge of 10^4 W1
P. knowlesi parasites (immature trophozoites) in a splenectomized
control (□——□), a normal control (■——■) and 2 rhesus monkeys
previously splenectomized and vaccinated with W1 merozoites in FCA:
(o——o) MZ65 and (●——●) MZ66.

TABLE II

VACCINATION OF RHESUS MONKEYS WITH P. KNOWLESI (W1) MEROZOITES
(PREPARED BY PHA OR CELL SIEVE (CS) AND SUBSEQUENTLY
FREEZE-DRIED (FD) OR FORMOL TREATED AND FREEZE-DRIED (F/FD))
RESULTS OF CHALLENGE INFECTIONS

Monkey	Vaccination		Challenges		Parasitaemias		
	Prep.	Variant	Variant	Day[1]	Prepatent period (days)	duration of patency (days)	Maximum parasitaemia (per 10^4 rbc)
MZ64[3]	PHA	W1	W3	56	6	13	350
	FD		W1	161	5	7	260
			W1	315	6	7	620
			W1	372	–	–	–[2]
BW276	PHA	W1	W1	63	10	2	0.01
			W1	101	5	10	20
			W1	223	5	18	240
MZ22	PHA	W1	W1	63	8	11	22
			W3	85	–	–	–
			W1	231	6	2	20
MZ41	CS	W1	W1	58	6	12	320
	F/FD		W3	60	6	4	3
MZ42	CS	W1	W3	80	3	10	190
	F/FD		W1	113	–	–	–
			W3	142	5	1	0.01
BW283	PHA	W1	W1	63	10	1	0.02
			W1	101	–	–	–
			W1	223	7	12	70

(1) Indicates days after vaccination

(2) Indicates that no parasites were observed by thick or thin blood film analysis

(3) In addition to the challenges shown two spleen biopsies were performed (removing approximately 1/10 of the spleen) on days 175 and 399.

TABLE III

SPLENECTOMY AND CHALLENGE OF PREVIOUSLY-VACCINATED,
IMMUNE RHESUS MONKEYS (SEE TABLE II)

Monkey	Day of splenectomy[1]	Challenges		Parasitaemias				Inhibitory A	
		Day[1]	Variant	Prepatent period (days)	Duration of patency (days)	Maximum parasitaemia (per 10^4 rbc)	Day of death (post challenge)	% (day)[1] antiW1	antiW
MZ64		517	W1[2]	0	11	3100	11	78(399)	
	518								
BW276	291	291	W1	5	5	1500	10	58(271)	
								34(291)	
MZ22	369	369	W1	11	1	0.01		69(298)	19(298
		402	W3	5	7	3000	12	62(400)	35(400
								53(411)	15(410
MZ41	170								
		196	W1	5	5	3000	10	65(196)	
								67(204)	
MZ42	259	259	W3	5	7	540		10(259)	53(259
		289	W1			negative		38(289)	42(272
		304	W1			negative		36(301)	
		315	W3			negative		69(324)	
BW283	291	291	W1	7	1	0.01		70(271)	
		354	W1	3	11	5300		33(291)	
								33(364)	

(1) Indicates days after vaccination

(2) MZ64 received 5×10^9 parasites 1 day prior to splenectomy giving an immediate paten parasitaemia of $.5/10^4$. This was cleared to subpatent levels within 12 hours but recrudesced after a further 2 days.

after splenectomy and remained resistant. Two monkeys (Table III – MZ22 and MZ41) died within 12 days of late challenge with prepatent periods and courses of parasitaemia similar to those observed in infected control animals. Inhibitory antibody levels against W1 or W3 variants were maintained after splenectomy in animals that died as well as in animals that survived late challenge infections. (Table III).

Splenectomy and challenge of immune marmosets

The results of challenge infections with P. knowlesi in four marmosets are summarized in Table IV. Three marmosets, initially challenged with W1 or W3 variants of P. knowlesi, recovered from this primary infection and were subsequently resistant to challenge with homologous or heterologous variants and strains of P. knowlesi. This resistance was maintained for intervals of up to 100 days between challenge infections. One animal (MZC9) was vaccinated with freeze-dried W3 merozoites of P. knowlesi in FCA, and challenged with an homologous variant of P. knowlesi. Parasites were observed in the peripheral blood within five days and persisted for 17 days with a maximum parasitaemia of 3.6%. MZC9 was subsequently challenged with a heterologous variant of the parasite (W1). No patent parasitaemia was observed. These marmosets were splenectomized 26 to 63 weeks after primary infection or vaccination (Table V). Challenge infections initiated on the day of splenectomy and up to 25 weeks after splenectomy produced negligible or low grade parasitaemias of brief duration. Challenge of one animal (C3) with a heterologous strain (Nuri) 26 weeks after splenectomy resulted in a patent infection within 12 days which persisted for 12 days and with a maximum parasitaemia of 16%. This animal subsequently recovered and was resistant to further challenge with P. knowlesi 37 weeks after splenectomy. No inhibitory antibody was detected in marmosets C3, C11 and C12. However, inhibitory antibody was detected in the serum of MZC9 prior to and after splenectomy (Tables IV and V).

Effects of immune spleen cells on the in vitro multiplication of P. knowlesi

The in vitro parasite levels of W1 P. knowlesi were determined after 24 hours' incubation in the presence of immune spleen cells and normal spleen or peripheral blood mononuclear cells. The details of the monkey donors of spleen cells are summarized in Table VI, and the characterization of the cells obtained from them are shown in Table VII. The starting parasitaemias and the parasitaemias recorded after 24 hours at effector to target ratios of 5:1, 2.5:1 and 1:1 are shown in Fig. 2. Logarithmic multiplication

TABLE IV   RESULTS OF P. KNOWLESI CHALLENGE INFECTIONS IN FOUR MARMOSETS

Marmoset	Challenge or vaccination		Parasitaemias			Inhibitory antibody (day)(1)	
	Variant or strain	Day (1)	Prepatent period (days)	Duration (days)	Maximum parasitaemia (per 10^4 rbc)	anti W1	anti W3
C3	W3	1	10	15	3500		0(47)
	W3	126	4	2	1		
C11	W1	1	5	15	4800		
	W3	78	8	4	5		
	W1	217	9	1	1	0(235)	
	Nuri	316	–	–	–	0(314)	
C12	W3	1		6	1800		
	W3	43	–	–	–		
	W1	182	–	–	–	0(180)	
	Nuri	210	–	–	–		
MZC9	W3MZ(2)	1,15,30					
	W3	106	4	17	360	37(76)	
	W1	235	8	1	1		

(1) Day after initial infection

(2) Vaccinated with 2×10^9 W3 freeze-dried merozoites in FCA on days shown.

TABLE V

SPLENECTOMY AND CHALLENGE OF IMMUNE MARMOSETS (SEE TABLE IV)

Marmoset	Splenectomy Day(1)	Challenge		Parasitaemias			Inhibitory antibody titres	
		Day(1)	Variant/ strain	Prepatent period (days)	Duration of patency (days)	Maximum parasitaemia (per 10^4 rbc)	anti W1 %	(day)(1)
C3	181	181	W3	6	5	1		
		265	W1	-	-	-	0	(284)
		363	Nuri	12	12	1600		
		483	W(2)	-	-	-		
C11	443	443	W(2)	-	-	-	0	(314)
C12	280	280	W1	-	-	-	0	(321)
		401	W(2)	-	-	-		
MZC9	306	306	W3	8	1	1		
		353	W1	-	-	-	52	(339)
		459	W(2)	-	-	-		

(1) Days after primary infection or vaccination

(2) Variant of W strain not determined.

TABLE VI

HISTORY OF RHESUS AND KRA MONKEY SPLEEN CELL DONORS. (SEE TABLE VII)

Monkey	Species/ source	Merozoite vaccination		Challenges		Splenectomy (day)
		Strain	Adjuvant	Variant	Day	
BW276				W1	63	
	rhesus/ imported	W1	FCA	W1	101	
BW283				W1	223	291$^{(1)}$
F28	kra/U.K.	–	–	W1	1	35$^{(2)}$
F37	kra/ malaysia	–	–	(4)		18$^{(3)}$

(1) Indicates days after primary vaccination
(2) Indicates days after infection
(3) Indicates days after drug treatment
(4) Infected with Plasmodium sp. on arrival. Drug cured

TABLE VII

IDENTIFICATION OF CELLS USED FOR IN VITRO CULTURE WITH P. KNOWLESI-PARASITIZED ERYTHROCYTES (SEE FIG. 2 and TABLE VI)

Monkey	Species	Source of cells	Polymorphonuclear cells %	Mononuclear cells %			
				Total	E rosettes +ve	Surface Ig	Monocytes
F28	kra	spleen	5	95	23	24	10
F37	kra	spleen	10	90	24	19	9
R1	kra	peripheral blood	4	96	46	10	2
BW276	rhesus	spleen	11	86	41	15	12
BW283	rhesus	spleen	9	91	44	21	9
G580	rhesus	spleen	3	97	26	25	8
G585	rhesus	peripheral blood	22	78	47	14	4

FIGURE 2

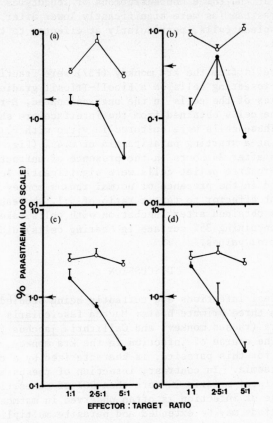

Wl P. knowlesi parasitaemias after 24 hours in vitro
incubation with:

a) Infected kra monkey spleen cells F28 (●——●) and normal kra
peripheral blood cells (o——o)
b) Infected kra monkey spleen cells F37 (●——●) and normal rhesus
monkey peripheral blood cells (o——o)
c) Merozoite-vaccinated rhesus monkey spleen cells BW276 (●——●)
and normal rhesus monkey spleen cells (o——o)
d) Merozoite-vaccinated rhesus monkey spleen cells BW283 (●——●)
and normal rhesus monkey spleen cells (o——o)

The horizontal arrows on the ordinate represent the starting
parasitaemias for each culture, the vertical bars represent the
standard errors of duplicate counts.

of normal spleen or peripheral blood cells 1-6 fold multiplication
was observed at the three concentrations of leucocytes used. In
contrast, parasitaemias were significantly lower after culture
with immune spleen cells, particularly at effector to target
ratios of 5:1.

Spleen cells from the kra monkey (F37) were fractionated by
removal of E-rosetting cells on a Ficoll-Triosil gradient. The
characteristics of the cells in the unfractionated, E-rosette
pellet and the cells obtained from the interface are shown in
Table VIII. These cells were cultured in vitro with P. knowlesi
Wl parasites at a starting parasitaemia of 0.45% (Fig. 3). The
parasitaemias after 24 hours in the presence of unfractionated
spleen and E-rosette pellet cells were significantly lower than
those obtained in the presence of normal rhesus monkey peripheral
blood cells at effector to target ratio of 5:1, whereas
parasitaemias obtained after incubation with cells taken from the
interface (containing 35% surface Ig bearing cells) did not differ
from the control values.

DISCUSSION

P. knowlesi infections are currently being studied in this
laboratory in three primate hosts: Macaca fascicularis (kra monkey),
Macaca mulatta (rhesus monkey) and Callithrix jacchus (common
marmoset). The course of infection in the kra monkey, which is the
natural host for this parasite, is characterized by a chronic low-
grade parasitaemia. In contrast, infection of rhesus monkeys result
in a rapidly fatal malaria in more than 90% of cases.(14) A
similar course of infection is often observed in marmosets, although
prepatent periods may be extended and parasite multiplication may
not always be logarithmic. Occasionally marmosets challenged by
intramuscular inoculation recover from a primary infection and are
subsequently resistant to challenge with different variants and
strains of P. knowlesi.(15) Protective immunity induced by
vaccination with merozoites in FCA has been extensively studied in
the rhesus monkey.(1-3) Preliminary investigations demonstrate
that such vaccination is also effective in kra monkeys (1) and
marmosets. Observations on the function of the spleen in
P. knowlesi infections of these three distinctive hosts are
preliminary and incomplete.

The spleen appears to be a necessary effector organ for the
production of a satisfactory primary immune response.(16-17)
Primary antibody responses can be elicited in the absence of a
spleen only when antigen is administered in sufficiently high

TABLE VIII

FRACTIONATION OF F37 SPLEEN CELLS: IDENTIFICATION OF CELLS USED FOR
IN VITRO CULTURE OF P. KNOWLESI (SEE FIG. 3)

Cells	Polymorpho-nuclear cells %	Cell characteristics Mononuclear cells %				
		Total	'E' rosettes	Surface Ig	Monocytes	"null"
Unfractionated	10	90	24	19	9	28
'E' rosette pellet	24	76	54	4	2	16
'E' rosette interface	2	98	3	35	6	54

220

FIGURE 3

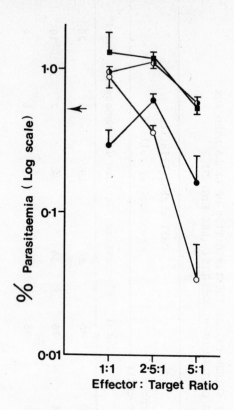

Parasitaemias of Wl P. knowlesi after 24 hours incubation
in vitro with normal rhesus monkey peripheral blood leucocytes
(▢——▢), unfractionated F37 spleen cells (●——●), E-rosette
pellet (○——○) and E-rosette depleted (interface) cells (●——●).
Other symbols are as in Fig. 2.

concentrations to stimulate specific cells in the peripheral
lymphoid organs.(17)   The extent of splenic involvement in the
primary immune response elicited by blood infection with P. knowlesi
is likely to differ from that elicited by vaccination, since in the
latter merozoites are administered intramuscularly in the presence
of FCA.

The role of the spleen in the induction of immunity to malaria has so far been investigated only in merozoite vaccinated rhesus monkeys. In contrast to intact animals vaccination did not protect previously splenectomized monkeys against homologous challenge with P. knowlesi. A partially-effective immune response was elicited in the absence of a spleen, as shown by the production of specific inhibitory antibody, and the abnormal course of parasitaemia. Although the parasite variant responsible for terminal parasitaemia was not typed, previous observations have suggested that similar patterns of parasitaemia may be associated with antigenic change of the parasite (G.H. Mitchell and G.A. Butcher, unpublished observation). There is evidence that intact vaccinated rhesus can mount a secondary type antibody to new challenge variants.(4) The later fulminating infection observed in the splenectomized monkeys may be due to the inability of the host to mount a sufficiently rapid antibody response to new parasite variants. Alternatively, the susceptibility of merozoite vaccinated, splenectomized monkeys despite significant specific antibody production, suggests that the spleen may provide an essential source of non-antibody producing cells.

The effector role of the spleen in immunity to P. knowlesi was investigated in merozoite-vaccinated rhesus monkeys and in previously infected, immune marmosets. The results in the former show that the role of the spleen differs among individual animals. When challenged at the time of splenectomy, two animals died of malaria and three survived. When four vaccinated rhesus monkeys were challenged 4-6 weeks after splenectomy, two suffered fatal infections and two were immune.

These results support the observations that inhibitory antibody alone is not responsible for the protective response elicited by vaccination since all the animals examined had significant levels of inhibitory antibody against the challenge variant. The varying responses of these animals may be attributable to the numbers of specifically competent cells available in the peripheral lymphoid organs.

In contrast to these variable results in the rhesus, all four marmosets immunized by infection remained resistant to challenge for up to 46 weeks after splenectomy. It is clear that the spleen does not play an essential effector role in the protective immune response of these animals to P. knowlesi. Blood infection of marmosets, in contrast to what is observed in rhesus monkeys, does not appear to induce merozoite inhibitory antibody; mechanisms of immunity in these animals are currently being investigated.

Marmosets provide an interesting and convenient model for the study of immunity to P. knowlesi, since in vivo cellular responses can be investigated in chimaeric twins by adoptive transfer experiments. Preliminary results have suggested that adoptive transfer of immune spleen cells from a merozoite immunized marmoset into a twinned recipient will confer a substantial degree of immunity. (J.Langhorne, unpublished observation)

Since rhesus and kra monkeys are outbred, cellular mechanisms involved in the immune response induced by merozoite vaccination or by blood challenge cannot be investigated in vivo. Therefore there is a need to develop in vitro assays in which viable parasites are maintained in vitro for 24-36 hours, in the presence of lymphoid cells. P. knowlesi will grow and multiply in the presence of normal leucocytes, provided that the parasitaemia is low, and leucocyte to parasite ratios do not exceed 10:1.(J. Langhorne, unpublished observation) In addition to the technical difficulties of maintaining parasites in vitro, there is also the problem of monitoring target cell destruction. The 24-hour multiplication cycle of P. knowlesi precludes measurement of parasite death by release of ^{51}chromium from pre-labelled cells in assays with long incubation times, since the chromium label would be released on schizont rupture. Additionally the low parasitaemias necessary for the maintenance of viable parasites in vitro (1%) would render any specific release of radio-label insignificant, unless parasites and not uninfected erythrocytes could be selectively pre-labelled. We are currently investigating means of monitoring intracellular parasite death in vitro.

It was considered worthwhile initially to determine whether immune spleen cells inhibited the multiplication of P. knowlesi in vitro, by measuring parasitaemia after 24 hours' incubation. The preliminary experiments reported here suggest that immune spleen cells from vaccinated rhesus monkeys and infected kra monkeys do inhibit multiplication of the parasite. This inhibition could be due to a variety of factors. In the experiment with fractionated kra spleen cells inhibition by specific antibody secreted from plasma cells appears unlikely as the fraction enriched for B lymphocytes was not effective in reducing parasite numbers. Inhibition was associated with the fraction relatively enriched for E-rosetting cells and polymorphonuclear cells. The reduction of parasitaemias may have been due to the E-rosetting or null cells present, or due to phagocytosis by polymorphonuclear cells. Spleen cells taken from an immune animal presumably contain activated cells, and the loss of parasites could alternatively be explained by their increased utilization of essential nutrients, or the production of toxic by-products secreted into the medium. These possibilities are currently being investigated.

In conclusion, the spleen appears to be a necessary organ for the induction of a protective immune response elicited by merozoite vaccination of rhesus monkeys. It may not, however, have a vital effector role in the maintenance of immunity in these monkeys or marmosets. In vivo experiments and preliminary in vitro experiments support the view that cell-mediated mechanisms have an important role in immunity to malaria produced by vaccination and natural infection.

SUMMARY

P. knowlesi produces distinctive patterns of infection in three primate hosts, Macaca fascicularis (kra monkey), Macaca mulatta (rhesus monkey) and Callithrix jacchus (common marmoset). It appears that all three species can be effectively vaccinated with P. knowlesi merozoites in FCA. Preliminary studies on the role of the spleen in acquired immunity to malaria in these three species have shown:

(i) In the rhesus monkey the spleen is required for induction of effective immunity by merozoite vaccination, but its effector role differs among individual animals as shown by variable resistance to challenge following splenectomy of previously vaccinated, immune animals.

(ii) In the marmoset the spleen does not have an essential effector role since animals immunized by infection remain resistant to challenge for at least 11 months after splenectomy.

(iii) Resistance or susceptibility in splenectomized rhesus monkeys and marmosets cannot be correlated with levels of merozoite inhibitory antibody, emphasising the importance of non-antibody forming cells in acquired immunity to malaria. Preliminary in vivo observations suggest that adoptive transfer of spleen cells from a merozoite-immunized marmoset can confer immunity to a chimaeric twin. Spleen cells from infected or vaccinated kra monkeys inhibit P. knowlesi multiplication in vitro; this effect is not shown by a B-cell enriched fraction, but is retained by a fraction enriched for E-rosetting and polymorphonuclear cells. Further analysis of these phenomena may help to elucidate the nature of cell-mediated immunity in malaria.

REFERENCES

1. MITCHELL, G.H. ET AL. Merozoite vaccination against Plasmodium knowlesi malaria. Immunology, 29: 397 (1975).

2. RICHARDS, W.H.G. ET AL. Merozoite vaccination of rhesus monkeys against Plasmodium knowlesi malaria; immunity to sporozoite (mosquito-transmitted) challenge. Parasitology, 74: 191 (1977).

3. MITCHELL, G.H. ET AL. A freeze-dried merozoite vaccine effective against Plasmodium knowlesi malaria. Clinical and Experimental Immunology, 28: 276 (1977).

4. BUTCHER, G.A. ET AL. Antibody mediated mechanisms of immunity to malaria induced by vaccination with Plasmodium knowlesi merozoites. Immunology, 34: 77 (1978).

5. TALIAFERRO, W.H. and MULLIGAN, A.W. The histopathology of malaria, with special reference to the function and origin of the macrophages in defense. Indian Medical Research Memoirs, 29: 1 (1937).

6. PHILLIPS, R.S. Plasmodium berghei; Passive transfer of immunity by antisera and cells. Experimental Parasitology, 27: 479 (1970).

7. DENNIS, E.D. ET AL. In vitro isolation of Plasmodium knowlesi merozoites using polycarbonate sieves. Parasitology, 71: 475 (1975).

8. COHEN, S. ET AL. Action of malarial antibody in vitro. Nature, 223: 368 (1969).

9. COHEN, S. and BUTCHER, G.A. Properties of protective malaria antibody. Immunology, 19: 369 (1970).

10. BÖYUM, A. Isolation of mononuclear cells and granulocytes from human blood. Scandinavian Journal of Clinical and Laboratory Investigation, 21: supplement 97: 77 (1968).

11. GREAVES, M.F. and BROWN, G. Purification of Human T and B lymphocytes. Journal of Immunology, 112: 420 (1974).

12. YAM, L.T. ET AL. Cytochemical identification of monocytes and granulocytes. American Journal of Clinical Pathology, 55: 283 (1971).

13. JANOSSY, G. ET AL. The effects of purified mitogenic proteins (Pa-1 and Pa-2) from pokeweed on human T and B lymphocytes in vitro. Clinical and Experimental Immunology, 26: 108 (1976).

14. GARNHAM, P.C.C. Malaria Parasites and Other Haemosporidia. Blackwell Scientific Press, Oxford (1966).

15. LANGHORNE, J. and COHEN, S. Plasmodium knowlesi in the marmoset (Callithrix jacchus). Parasitology, In press.

16. CAMPBELL, P.A. and LA VIE, M.F. Effect of splenectomy on primary and secondary response to sheep erythrocytes in rats. Proceedings of the Society of Experimental Biology and Medicine, 124: 571 (1967).

17. ROWLEY, D.A. The effects of splenectomy on the formation of circulating antibody in the adult male albino rat. Journal of Immunology, 64: 289 (1950).

ACKNOWLEDGEMENTS

This work is supported by the Medical Research Council, United Kingdom, and the World Health Organization. Rhesus monkeys BW276 and BW283 were part of collaborative experiments carried out with Dr W.H.G. Richards, Wellcome Laboratories of Tropical Medicine, Beckenham, Kent. We would like to thank Miss Marilyn Burnikell and Miss Deborah Cutter for skilled technical assistance.

DISCUSSION

MORRIS: This is a general question but the answer is of some relevance to a discussion on splenectomy in these various animal models of malaria. There must have been a considerable number of splenectomies performed for trauma in people with malaria – what happens to them? Do they die of malaria? Do they get exacerbations of their malaria?

McGREGOR: Published evidence indicates that exacerbation of clinical malaria frequently follows splenectomy in naturally-immunized persons. I have no personal experience of the subject.

DAVID-WEST: There is general agreement that splenectomy predisposes to malaria infection. The experience in Nigeria is also in agreement with Dr McGregor. The only case of cerebral malaria in an adult seen in the post-mortem records at University College Hospital, Ibadan, was in a man who had had a splenectomy for trauma some months previously.

McGREGOR: Published evidence indicates that splenectomy in naturally-immunized humans is often followed by exacerbation of parasitaemia.

CRANE: Of course, everyone has been aware of the potential risks of fatal malaria following splenectomy in man for a long time and certainly in Papua New Guinea it has been standard practice to give regular anti-malarials after splenectomy for any reason. When such patients eventually cease their anti-malarials and later experience a severe or fatal attack of malaria, it would be impossible to say whether this resulted from the splenectomy as such, or from the loss of immunity through a long period of protection following the operation.

COHEN: The antigenicity of the parasitized red-cell membrane could be a basic determinant of the role of the spleen in malaria immunity. It is worth mentioning, therefore, that our knowledge of this is extremely restricted. Variant-specific antigens are present in rhesus red-cell membranes parasitized by Plasmodium knowlesi and the knob-like protrusions of Plasmodium falciparum parasitized human red cells appear to be antigenic. In general, red cells containing early developmental forms of the parasite do not appear to express malarial antigens and this may account for the apparent failure of the spleen to clear parasites referred to by Dr Oster, especially as mature intracellular parasites are frequently immobilized in tissue capillaries and therefore excluded from the splenic circulation.

McGREGOR: A feature of <u>Plasmodium falciparum</u> malaria in humans is that, as intraerythrocytic parasites grow and mature, they withdraw from the general circulation and complete their development within the capillaries of internal organs. This phenomenon may favour parasite survival as it may greatly reduce the number of parasitized erythrocytes passing through the splenic cords at a time when the physical size of the parasite would make it singularly susceptible to trapping and destruction by the splenic mechanisms described by Professor Weiss and Dr Oster.

MITCHELL: Dr Langhorne, you have shown that splenectomy affects the efficacy of vaccination without affecting titres of merozoite-invasion-inhibition antibodies, is there any class difference in such antibody in splenectomized versus intact monkeys which may contribute to <u>in vivo</u> differences in behaviour? If not, the conclusion seems inescapable that inhibitory antibody is <u>insufficient</u> for species-specific host protection following vaccination.

LANGHORNE: Although immunity to homologous challenge can be passively transferred with serum from merozoite-vaccinated animals, there are several lines of evidence that indicate the importance of other specific immune mechanisms, as yet unidentified. The possible class differences in inhibitory antibody before and after splenectomy have not been studied.

NOGUEIRA: Dr Oster, for the experiments testing suppressor activity did you use splenic macrophages, that is, adherent cells from a spleen cell suspension? Did you use any markers to identify this population like, for instance, binding of C-coated erythrocytes?

OSTER: Adherent spleen cells were used in the results described here. Additionally, similar results were obtained when peritoneal adherent cells were employed in the lymphocyte activating factor assay. The adherent cells were identified by morphology; no additional markers were used in this study.

ZEMBALA: Is there any information concerning the nature of macrophage-suppressor factor(s)?

OSTER: The immunosuppressive factor also reduces the proliferative responses to spleen cells to PHA and malaria antigen, but not to the B-cell mitogen, LPS. It also diminished the responses to other antigen, such as PPD, indicating that this factor diminished T cell but not B cell proliferation. The immunosuppressive factor has been partially characterized. Lysozyme and interferon were assayed directly in the suppressive supernatants, and were not found in significant quantities. The inhibitory activity was unaffected by heat or neuroaminidase. However, trypsin, pronase and ribonuclease A abolished suppressive activity. Additional, 10^{-5} N indomethacin added at the onset of incubation of the adherent cells did not inhibit production of this factor, indicating that it is not a prostaglandin.

The Tropical Splenomegaly Syndrome

A review of its pathogenesis

B.M. Greenwood and Y.M. Fakunle,
Ahmadu Bello University, Zaria, Nigeria.

INTRODUCTION

Patients with massive splenomegaly are encountered more
frequently in the tropics than in countries with a temperate climate.
In some of these patients an obvious reason for the splenomegaly is
readily demonstrated but in many no infectious, haematological,
metabolic or neoplastic cause for the splenomegaly can be found.
Many of this heterogeneous group of patients have sufficient
clinical and laboratory features in common to justify the
definition of a distinct syndrome now usually called the tropical
splenomegaly syndrome (TSS). Even in tropical areas not all
patients with idiopathic splenomegaly have TSS. A recent study (1)
in Zaria, northern Nigeria of 75 patients over the age of seven
with massive splenomegaly showed that 28 had a definite cause for
their enlarged spleen. Thirty of the remaining 47 patients were
thought to have TSS but no diagnosis could be established in the
rest. We have subsequently found that, after full laboratory
investigation, a diagnosis of TSS cannot be sustained in nearly half
the patients in whom this diagnosis has been made on clinical
grounds nor can any other cause for their splenomegaly be found. It
seems likely that in many areas of the tropics patients are regularly
seen with idiopathic splenomegaly which cannot be attributed to TSS.
In contrast to patients with TSS, this group of patients has
received little attention from research workers.

The clinical features of patients with TSS have been fully reviewed on many occasions, most recently in the comprehensive survey of Marsden and Crane,(2) and will be considered only briefly in this paper which concentrates on current views on the pathogenesis of this syndrome.

DIAGNOSTIC CRITERIA

There is now general agreement over the main criteria needed to make a diagnosis of TSS but still some difference of opinion as to the relative weight which should be attached to the presence or absence of different features of the syndrome. Our diagnostic criteria (Table I) are fairly strict. We consider 1,2,3 and 6

TABLE I

CRITERIA FOR A DIAGNOSIS OF TSS

1. Massive splenomegaly ($>$ 15 cm) for which no other cause can be found.

2. Immunity to malaria.

3. Serum IgM at least two standard deviations above local mean.

4. Hepatic sinusoidal lymphocytosis.

5. Normal lymphocyte response to phytohaemagglutinin.

6. Clinical and immunological response to malarial prophylaxis.

major criteria essential for a diagnosis; 4 and 5 are usually, but not invariably, found.

The following are some points that should be considered in the application of these criteria.

1. Massive splenomegaly for which no other cause can be found

It is essential that strenuous efforts be made to direct any infectious or neoplastic cause for the patient's splenomegaly - the nature of the tests employed being determined partly by the local pattern of disease. Chronic infections such as brucellosis are easily missed. Fever is not a usual feature of TSS and its presence should raise the suspicion of an incorrect diagnosis.

2. Immunity to malaria

In an area of high malarial endemicity survival to adult life can usually be considered adequate evidence for the presence of malarial immunity. Inclusion of this criterion allows differentiation of TSS from the splenomegaly, which may be marked, which accompanies recurrent attacks of acute malaria. We have sometimes had diagnostic difficulties in older children, on the verge of acquiring effective malarial immunity, who have presented with features strongly suggesting a diagnosis of TSS. Measurement of malarial antibody levels might, perhaps, be helpful in these cases.

3. Raised serum IgM

Definition of a raised serum IgM will depend upon the local normal value. This is likely to be high because of the occurrence of TSS in areas of high malarial endemicity. At Zaria we require a value of over 1,000 IU/ml ($>$ 2 SD above the local normal mean) before accepting a diagnosis of TSS.

4. Hepatic sinusoidal lymphocytosis (HSL)

This is found in most, but not all, patients who satisfy all the other criteria for a diagnosis of TSS. It is not specific for TSS and may occur in other conditions such as Felty's syndrome and chronic lymphatic leukaemia.

5. Normal PHA response

Whilst agreeing with Sagoe (3) that patients with TSS normally have a good PHA response, in contrast to patients with a lymphoma, we have occasionally encountered patients with otherwise typical TSS who have a poor PHA response and we do not, therefore, consider this to be an essential diagnostic criterion. A poor PHA response is occasionally encountered in apparently healthy subjects.

6. Response to anti-malarials

If given regular malarial prophylaxis patients with TSS show a progressive clinical and immunological improvement which is sustained for as long as treatment is continued. However it is sometimes not appreciated that patients with other conditions, such as chronic lymphatic leukaemia, may show a reduction in spleen size when given malarial prophylaxis, probably because they have lost some of their immunity to malaria. However in this latter group of patients the reduction in spleen size is not usually marked and is often only transitory.

CLINICAL FEATURES

The tropical splenomegaly syndrome is seen most frequently in young and middle aged adults. At Zaria it is encountered more frequently in females than in males. The dominant clinical feature is massive enlargement of the spleen which may produce sufficient abdominal discomfort to make the patient go to hospital. The liver is often enlarged but lymphadenopathy is not a feature of the syndrome.

Most patients have some degree of normocytic anaemia and this may occasionally be severe, especially during pregnancy. Acute haemolytic episodes may occur. Hypervolaemia, splenic pooling and enhanced red cell destruction are all thought to contribute to the pathogenisis of the anaemia.(4) Overproduction of immunoglobulin and albumin is thought to be the main cause of hypervolaemia.(5) Leucopaenia is often found and is usually accompanied by a relative, and sometimes an absolute, lymphocytosis. Lymphocytosis is also found in the bone marrow and the haematological findings may therefore strongly suggest a diagnosis of chronic lymphatic leukaemia. This form of TSS appears to occur more frequently in West Africa than elsewhere and it is possible that other regional variations in clinical features of the syndrome occur.

A surprising feature of TSS is its high mortality. A 50% mortality has been found in small groups of patients followed for approximately 5 years in both Uganda and New Guinea (6,7) and we have also observed a high mortality among our patients. Death is often sudden, suggesting an overwhelming infection but this has rarely been documented. It is not obvious why patients with TSS should be unduly susceptible to infections. Crane et al. (8) reported that patients with TSS seen in New Guinea had a diminished primary immune response to flagellin but in general cellular and humoral immune responses are well maintained (9, Fakunle - unpublished data). The possibility that TSS is a pre-malignant condition has been widely discussed but malignant transformation in a patient with TSS has never been clearly documented. Nevertheless it is possible that some deaths in patients with TSS are due to the development of a lymphoma. A long-term clinical and

immunological follow-up of a group of patients with TSS would
provide a great deal of valuable information but, unfortunately, in
many parts of the world where the condition occurs it is seen
predominantly among rural and nomadic people among whom follow-up
is very difficult.

TREATMENT

Regular malarial prophylaxis causes the spleen of patients
with TSS to shrink and their immunological abnormalities to regress.
Several months of treatment may be required before a clinical
response is obtained and improvement may continue for at least a
year. In a few patients the spleen never returns completely to
normal. If treatment is stopped splenomegaly and the accompanying
immunological abnormalities recur.(10) Although no long-term
follow-up studies have been reported the life expectancy of a
patient with TSS is probably normal if he continues with regular
prophylaxis.

Splenectomy leads to an immediate symptomatic improvement and
a fall in the serum IgM.(11) However removal of a massive spleen
carries a significant operative mortality and reduces the
resistance of the patient to acute infections such as pneumococcal
septicaemia and malaria. We believe that provided a diagnosis of
TSS has been firmly established splenectomy should not be carried
out in patients with this condition.

PATHOGENESIS

The pathogenesis of TSS has excited considerable interest, not
only among those concerned with the care of patients with this
condition, but also among immunologists in general as the unusual
immunological features of this condition may perhaps offer important
clues to the understanding of the mechanisms involved in the normal
immune response to malaria. Although the pathogenesis of TSS is
still not fully understood, we believe that sufficient information
has now been collected to allow the construction of a working
hypothesis which can account for many of the features of the
syndrome.(12,13)

The Hypothesis

The hypothesis (Fig. 1) suggests that patients with TSS have
an abnormal immune response to malaria, perhaps genetically
determined, which results in excessive IgM production and the
formation of large molecular weight immune complexes. These

FIGURE 1

A POSSIBLE SCHEME FOR THE PATHOGENESIS OF TSS

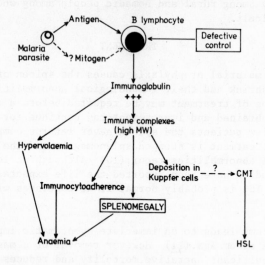

complexes are taken up by phagocytic cells of the liver and spleen and may adhere to red cells causing enhanced red cell destruction. Massive stimulation of the reticulo-endothelial system maintained over many years leads to splenomegaly. Sequestration of lymphocytes in the liver and spleen accounts for hepatic sinusoidal lymphocytosis and T lymphocytes found in the liver may be involved in a cell-mediated immune reaction to liver cell antigens.

The various components of this suggested pathway are now considered in turn.

Malaria and TSS

A critical role for malaria in the pathogenesis of TSS seems incontrovertible - some of the evidence is summarized in Table II. It was initially suggested that Plasmodium malariae might have a special role in producing TSS but this view has not been sustained and it is now thought that TSS can follow repeated infection with any of the human malaria parasites. The pattern of malaria infection may, however, be important in predisposing to the development of TSS. In New Guinea TSS is seen most frequently in areas where there is some seasonal variation in the transmission of malaria. In Uganda the condition is found frequently among

TABLE II

EVIDENCE LINKING MALARIA TO THE PATHOGENESIS OF TSS

1. Geographical distribution of the condition corresponds to that of malaria.

2. High malarial antibody levels in affected patients.

3. Rarity of the condition in those with the genotype AS.

4. Therapeutic effects of malarial prophylaxis.

5. Production of a TSS-like syndrome in experimental animals by malaria infection.

Rwandans who have migrated from an area of low malarial endemicity to one of high endemicity. In northern Nigeria the condition is seen most commonly among nomadic cattle Fulani whose exposure to malaria may be greater than that of the local settled population.

There is no evidence to suggest that patients with TSS have a diminished protective immune response to malaria and several observations suggest the reverse. Thus patients with TSS have fewer episodes of parasitaemia, less malarial pigment in their Kupffer cells and higher malarial antibody levels than control subjects living in the same area. The possible existence of a more subtle immune defect, such as a defect in cell-mediated immunity to malaria or an abnormal pattern of handling malarial antigens or mitogens has, however, yet to be excluded.

Macroglobulinaemia and TSS

Macroglobulinaemia has been recognized as a cardinal feature of TSS since the time that the syndrome was first described.(14) The serum level of IgM may be very high and values up to 50 times that of the local normal mean have been recorded. The mean serum IgM of a group of patients with TSS that we have recently studied was 3567 IU/ml compared with a mean value of 261 IU/ml in a group of healthy controls. Turn-over studies using I^{125} labelled IgM (15) have shown that these high serum levels are due to increased IgM production and not to a decreased rate of IgM catabolism (Table III).

Immunoelectrophoresis shows that the serum IgM is polyclonal; in only one patient have we found an IgM para protein. Only a small proportion of the macroglobulin can be shown to be malaria antibody,

TABLE III

IgM HALF-LIFE AND TURN-OVER TIMES IN FIVE PATIENTS
WITH TSS AND IN FIVE HEALTHY NIGERIAN CONTROLS

	Half-life hr	Turn-over time hr
Patients	42 + 14	30 + 19
Controls	31 + 7	17 + 6

Figures indicate the mean + 1 standard deviation.(15)

the remainder contains antibodies with specificity for altered IgG
and for other autoantigens. A small proportion of the large
amounts of IgM found in the serum of patients with TSS is low
molecular weight 7S IgM but the amount present, usually less than
5% is not sufficient to cause serious inaccuracies in determining
IgM levels in patients with this condition by the Mancini technique.
In addition to possessing low molecular weight IgM in the serum
many patients also have free light chains in the serum and urine
(16)(Table IV).

TABLE IV

THE NUMBER OF PATIENTS WITH TSS AND OF HEALTHY NIGERIAN AND
EUROPEAN CONTROLS WITH LOW MOLECULAR WEIGHT (7S) IgM AND
FREE LIGHT CHAINS IN THE SERUM.(16)

	7S IgM		Free light chains
	Polyacrylamide/ agarose diffusion	10% PEG precipitation	Immunodiffusion
TSS patients n.31	19	15	29
Nigerian controls n.28	1	2	0
European controls n.21	0	0	0

The macroglobulinaemia of TSS is accompanied by a marked increase in the proportion and total number of circulating B lymphocytes (17) (Table V). Similar results are obtained with both

TABLE V

PROPORTION AND ABSOLUTE NUMBER OF B LYMPHOCYTES IN PATIENTS
WITH TSS AND IN HEALTHY NIGERIAN CONTROLS

	Percentage	Cells/μl
Patients n. 39	39 + 12*	1,080 + 653*
Controls	22 + 4	488 + 180

Figures indicate mean + 1 standard deviation. (17)

*P = 0.001

immunofluorescence and the EAC rosette technique. Passive absorption of IgM onto lymphocytes is unlikely to be the explanation for the high B lymphocyte counts found in patients with TSS as similar results are found before and after incubation in tissue culture medium. On treatment B cell counts fall to within the normal range.

·These findings suggest that in patients with TSS there is marked polyclonal B lymphocyte stimulation with massive, and partially uncontrolled, immunoglobulin production.

Immune complexes and TSS

Very high cryoglobulin levels are found in the serum of most patients with TSS. (18) In a recent study we found a mean level of 897 + 571 μg/ml in 13 patients with TSS compared with a mean level of 27 + 15 μg/ml in 15 healthy controls. These cryoglobulins contain IgM, IgG and sometimes C3 but have not been shown to contain malarial antibody or antigen. Using more refined methods for the detection of immune complexes we have recently demonstrated the presence of very high levels in the serum of most patients with TSS (19) (Fig. 2). These complexes contain large amounts of IgM and it is possible that they are in part aggregates of IgM behaving biologically in the same way as true antigen-antibody immune complexes. The presence of immune complexes on the red cells of patients with TSS has also been reported.

238

FIGURE 2

IMMUNE COMPLEX LEVELS ASSESSED BY THREE DIFFERENT METHODS IN
SERA FROM 13 PATIENTS WITH TSS.   THE HATCHED AREA
INDICATES THE NORMAL SERUM RANGE (mean ± 2 STANDARD DEVIATIONS)
FOUND IN HEALTHY NIGERIANS. (19)

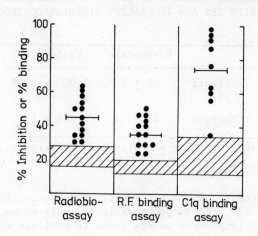

Large amounts of immunoglobulin, probably complexes, can be
shown in the Kupffer cells of patients with TSS by immunofluorescence
and IgM is present in macrophages in the spleen. Some idea of the
load imposed on the phagocytic cells of the reticulo-endothelial
system in patients with this condition can perhaps be gained by
the simple experiment of incubating normal polymorphonuclear
neutrophils in serum from a patient with TSS. Within 30 minutes
the leucocytes can be shown by immunofluorescence to have picked up
numerous lumps of IgM. We believe that an attempt to clear these
complexes is the main reason for the splenomegaly found in TSS.
Support for this view comes from the observation that marked
enlargement of the liver may occur in patients who have had a
splenectomy if they do not continue with malarial prophylaxis.

Hepatic sinusoidal lymphocytosis (HSL)

Although not specific for TSS, hepatic sinusoidal lymphocytosis
is a regular and puzzling feature of the condition. We have
recently shown that most of the lymphocytes infiltrating the

hepatic sinusoids are T lymphocytes (unpublished data) suggesting that a cell-mediated immune reaction may be occurring at this site. Adherence of lymphocytes to Kupffer cells has been observed(20) suggesting that a cell-mediated immune reaction to a component of the complexes present in these cells might be taking place. We therefore studied the response of patients with TSS to cryoglobulins, obtained from their serum, in lymphocyte transformation and leucocyte migration inhibition tests. Cell-mediated immunity to this material could not be demonstrated. In a further series of experiments (unpublished data) leucocyte migration tests were carried out using an extract of a liver biopsy specimen obtained from a patient with TSS. Significant inhibition of migration was observed with leucocytes obtained from patients with TSS but not with leucocytes obtained from controls (Fig.3). Inhibition was also observed with an extract of normal

FIGURE 3

MIGRATION OF LEUCOCYTES OBTAINED FROM PATIENTS WITH TSS AND FROM NORMAL CONTROLS IN THE PRESENCE OF LIVER EXTRACTS PREPARED FROM THE LIVER OF A PATIENT WITH TSS AND FROM NORMAL LIVER.

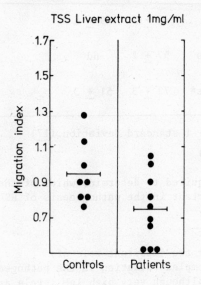
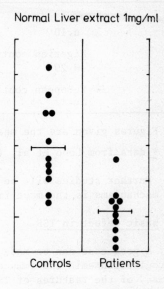

TSS Liver extract 1mg/ml

Normal Liver extract 1mg/ml

liver suggesting that sensitisation to a liver cell antigen had
occurred. This has been noted in patients with other forms of
liver disease but was a surprising finding in patients with TSS in
whom little evidence for liver cell damage has been found on light
or electron microscopy.(20)

Non-specific sequestration of T lymphocytes in the liver and
spleen is an alternative explanation for HSL for which there is
some supporting evidence as patients with TSS have a lower
proportion of circulating T lymphocytes than normal with an
associated increase in the proportion of T lymphocytes in splenic
aspirates (Table VI).

TABLE VI

PROPORTION OF T LYMPHOCYTES IN THE PERIPHERAL BLOOD AND IN
THE SPLEEN OF PATIENTS WITH TSS AND OF NIGERIAN AND
EUROPEAN CONTROLS

	Blood %	Spleen %
Patients n.10	43 + 5	49 + 6
Nigerian controls n.29	57 + 7	nd
European controls* n.11	72 + 3	51 + 3

Figures given are the mean + 1 standard deviation.(17)

* data from Lobo et al. (21)

Further studies will be required to determine which of these
two mechanisms is the most important in the pathogenesis of HSL.

The basic defect in TSS

A good case can be made for the important role of
macroglobulinaemia and immune complex formation in the pathogenesis
of many of the features of TSS although very high IgM levels are
occasionally found in patients with infections such as
trypanosomiasis, and even in apparently normal individuals, who have
no other features of the syndrome. Why patients with TSS have such

high serum IgM levels is, however, still completely unknown. This is not surprising as little is known about the pathogenesis of the hypergammaglobulinaemia that normally accompanies malaria.

During an attack of acute malaria the serum IgM rises rapidly to a level two or three times higher than normal. Only a small proportion of this immunoglobulin is malarial antibody and there is some evidence that a parasite-derived B-cell mitogen, acting in a manner analogous to bacterial endotoxin, may play some part in the production of the rest. Acute malaria produces marked changes in lymphocyte subpopulations and an imbalance in the cellular mechanisms controlling immunoglobulin synthesis may also be involved. Once an acute attack of malaria has been treated the serum IgM gradually returns to normal. However in those exposed repeatedly to malaria the serum IgM gradually rises over a period of many years levelling out at a value two or three times greater than that found in healthy Europeans. Some homeostatic mechanisms must come into operation at this point but their nature is unknown. We believe that the most likely explanation for the macroglobulinaemia of TSS is that, in patients with this condition, these homeostatic mechanisms are defective and only come into operation when very high serum IgM levels have been reached. The nature of this defect is completely speculative but defective function of T suppressor cells is a possibility.

Whatever its nature the basic immune defect in patients with TSS is likely to be genetic. Multiple cases are regularly found in the same family and a familial tendency to develop the disease can sometimes be very strong.(12) A small HLA study carried out in Zaria (unpublished data) has given inconclusive results but it is possible that the wrong genetic markers were being investigated; perhaps a study of B lymphocyte surface antigens would be rewarding.

CONCLUSION

In this paper we have presented a hypothesis which, we think, can explain many of the known clinical and laboratory features of TSS. There are however many other possible explanations for the abnormal immunological features of the syndrome - defective macrophage handling of malarial antigens, production of low affinity malarial antibody and defective cell-mediated immunity to malaria being only a few of the possibilities that are perhaps worthy of further study.

During the past few years new techniques have been established which allow the identification of sub-groups of cells involved in different functional aspects of the immune response. Application of these techniques to the study of human diseases, such as the immune deficiencies, has been very rewarding and is likely to be similarly rewarding in the study of patients with TSS. This will only be possible when laboratories with moderately sophisticated resources are established in areas where patients with TSS are found. Fortunately the number of such laboratories is increasing and it is likely that the next few years will see further major advances in our understanding of the pathogenesis of TSS.

SUMMARY

The criteria necessary for a diagnosis of the tropical splenomegaly syndrome (TSS) are discussed and the pathogenesis of the syndrome reviewed. It is suggested that TSS follows defective regulation of the macroglobulinaemia that normally accompanies malaria infection. High molecular weight immune complexes are formed in large amounts and taken up by phagocytic cells in the liver and spleen leading to hepatomegaly and splenomegaly. Infiltration of hepatic sinusoids with T lymphocytes may result from a cell-mediated immune reaction to a liver cell antigen or from non-specific sequestration of T lymphocytes in the liver and spleen. Laboratory techniques are now available which should allow the immune defect in patients with TSS to be more clearly defined.

ACKNOWLEDGEMENTS

We thank Professor A.F. Fleming and Dr H.C. Whittle for their helpful comments.

REFERENCES

1. BRYCESON, A.D.M. ET AL. Splenomegaly in northern Nigeria. Acta tropica, 33: 185 (1976).

2. MARSDEN, P.D. and CRANE, G.G. The tropical splenomegaly syndrome - a current reappraisal. Revista do Instituto de Medecina tropicale de Sao Paulo, 18: 54 (1976).

3. SAGOE, A-S. Tropical splenomegaly syndrome: long term proguanil therapy correlated with spleen size, serum IgM and lymphocyte transformation. British Medical Journal, 3: 378 (1970).

4. PRYOR, D.S. The mechanism of anaemia in tropical splenomegaly. Quarterly Journal of Medicine, 36: 337 (1967).

5. CRANE, G.G. ET AL. The role of plasma proteins in chronic expansion of plasma volume in tropical splenomegaly syndrome. I - The metabolism of I^{131} labelled albumin. Transactions of the Royal Society of Tropical Medicine and Hygiene, 68: 246 (1974).

6. HAMILTON, P.J.S. ET AL. Splenectomy in tropical splenomegaly syndrome - a five year follow up. Journal of Tropical Medicine and Hygiene, 74: 230 (1971).

7. CRANE, G.G. ET AL. Tropical splenomegaly in New Guinea. I. Natural history. Transactions of the Royal Society of Tropical Medicine and Hygiene, 66: 724 (1972).

8. CRANE, G.G. ET AL. Humoral immune responses in the tropical splenomegaly syndrome in New Guinea. Clinical Science, 43: 869 (1972).

9. ZIEGLER, J.L. ET AL. Immunological studies in tropical splenomegaly syndrome in Uganda. British Medical Journal, 4: 15 (1969).

10. DAVID-WEST, A.S. Relapses after withdrawal of proguanil treatment in tropical splenomegaly syndrome. British Medical Journal, 3: 499 (1974).

11. CRANE, G.G. ET AL. Tropical splenomegaly syndrome in New Guinea. II. Long term results of splenectomy. Transactions of the Royal Society of Tropical Medicine and Hygiene, 66: 733 (1972).

12. ZIEGLER, J.L. and STUIVER, P.C. Tropical splenomegaly syndrome in a Rwandan kindred in Uganda. British Medical Journal, 3: 79 (1972).

13. FAKUNLE, Y.M. and GREENWOOD, B.M. A suppressor T-cell defect in tropical splenomegaly syndrome. Lancet, 2: 608 (1976).

14. WELLS, J.V. Serum immunoglobulin levels in tropical splenomegaly syndrome in New Guinea. Clinical and Experimental Immunology, 3: 943 (1968).

15. FAKUNLE, Y.M. and GREENWOOD, B.M. Metabolism of IgM in the tropical splenomegaly syndrome. Transactions of the Royal Society of Tropical Medicine and Hygiene, 70: 346 (1976).

16. FAKUNLE, Y.M. and GREENWOOD, B.M. Low molecular weight (7S) IgM and free light chains in the sera of patients with the tropical splenomegaly syndrome. Clinical and Experimental Immunology, 28: 153 (1977).

244

17. FAKUNLE, Y.M. ET AL. T- and B- lymphocyte subpopulations in the tropical splenomegaly syndrome. Clinical and Experimental Immunology, in press.

18. ZIEGLER, J.L. Cryoglobulinaemia in the tropical splenomegaly syndrome. Clinical and Experimental Immunology, 15: 65 (1973).

19. FAKUNLE, Y.M. ET AL. Cryoglobulinaemia and circulating immune complexes in tropical splenomegaly syndrome. Clinical and Experimental Immunology, 31: 55 (1978).

20. FLUCK, D.J. ET AL. Ultrastructure of the hepatic sinusoids in the tropical splenomegaly syndrome. Transactions of the Royal Society of Tropical Medicine and Hygiene, 67: 638 (1973).

21. LOBO, P.I. ET AL. Kidney transplantation across a positive cross-match; cross-match assays and distribution of B lymphocytes in donor tissues. Lancet (i): 925 (1977).

The Serology of Tropical Splenomegaly Syndrome and its Relationship to Malaria

Dr G.G. Crane, Division of Haematology, Repatriation General
Hospital, Concord, NWS, Australia.

It is now generally accepted that tropical splenomegaly syndrome (TSS) arises through an abnormal immune response to recurrent malarial infection, and that there is no evidence of any additional extrinsic etiological agent.(1) None of the major characteristics of the syndrome - persistent splenomegaly in the absence of commensurate parasitaemic episodes, hepatic sinusoidal lymphocytosis (HSL), and polyclonal macroglobulinaemia - form part of the normal response to malaria. Together with the exceptionally high levels of immunofluorescent antimalarial antibody (IFA), they are all regarded as manifestations of the immunological aberration.

Recently, much attention has been paid to other serological phenomena in TSS - in particular to the prevalance of cryoglobulins, low complement (C3) levels, red-cell agglutinins in serum or on red cells, and other autoantibodies such as rheumatoid factor. However, it is not clear that all these findings occur exclusively, or even to a greater degree, in TSS, rather than being components of the normal response to malaria. This distinction is vital to any hypothesis seeking to explain the mechanisms whereby certain individuals respond to malaria in this abnormal way, and to elucidate the role of the spleen in the pathogenesis of this disease.

In attempting to document the specific deviations of TSS from the normal immune response to malaria, we have carried out a series of field and laboratory studies in Papua New Guinea over a number of years.(2) The syndrome is exceptionally common in a number of malarious areas of the country, particularly in the Upper Watut Valley where up to 95% of adult inhabitants of different villages are affected. Furthermore, no other cause of persistent splenomegaly occurs there with any frequency. It has therefore been possible to compare both this population as a whole with environmentally similar New Guinean populations in whom TSS is rare or non-existent, and also within this one tribal group to compare affected individuals with unaffected controls matched for age, sex and environment.

Our studies may be summarized as follows:

Relationship to malaria: TSS is found in some parts of the hyperendemic coast of Papua New Guinea, notably in the Sepik Province, but is most common in certain meso-endemic areas, 1,000 - 1,500 m. above sea level, adjoining the relatively malaria-free highlands. Such areas do not show a distinctive pattern of transmission of malaria, nor an abnormal distribution of plasmodial species, Plasmodium vivax predominating throughout the country. Subjects with TSS acquire active immunity to malaria normally, as judged by falling parasite rates with increasing age, but splenomegaly fails to regress. Adults with TSS have in fact significantly lower parasite rates than matched controls from the same villages.

Epidemiology: The distribution of TSS in Papua New Guinea could be equally well explained on a tribal basis as on a geographical or enivironmental one, and in the Upper Watut Valley is clearly familial. However, studies of red-cell antigens and enzymes, serum genetic markers and HLA A and B series antigens have failed to demonstrate any specific associations with TSS and in the Watuts, who are remarkable for their lack of HLA polymorphism.

Immunoglobulins: Total circulating IgG is significantly greater in TSS than in controls, but this increase is masked by the associated expansion of plasma volume. Consequently, both serum IgG and IgA concentrations are comparable in subjects with and without TSS, and remain steady throughout adult life. By contrast, serum IgM concentrations are markedly higher in TSS, continue to rise throughout life, and are highest in those with the biggest spleens. It is noteworthy that they are significantly elevated even in three and four year old Watut children, prior to the development of any other sign of TSS.

Malaria antibodies: IgG and IgM IFA titres to all species of plasmodia are significantly higher in TSS than in matched controls. Both continue to rise throughout life, this trend being significantly more marked in the case of IgM IFA. In normals, by contrast, peak titres are reached in adolescence.(3)

Cold agglutinins of anti-I, anti-i or of no definable specificity, in titres greater than 1:32 are found occasionally in children, but significantly more frequently in adults with or without TSS. High titres (greater than 1:256) occur only in adults with TSS, but neither the mean titre nor the proportion of subjects with elevated values is significantly greater in TSS than in controls (28% of 24%). This is also true of antibodies to enzyme-treated red cells, found in 52% of adults with TSS and in 46% of adult controls.

Cryoproteins occurred in 85% of adults with established TSS, but also in 25% of adult controls. Rheumatoid factor (by Latex flocculation) has been found in 87% of subjects with TSS, and in 35% of controls.

These studies demonstrate that the first detectable abnormality to appear is an elevation in serum IgM levels. Later, as active immunity develops and parasite loads fall, instead of regression of splenomegaly, transition occurs from simple malarious splenomegaly to TSS. Immunoglobulin and malariometric data indicate that this transition occurs between the ages of eight and 15 years and, although the appearance of HSL has not been documented extensively in this age group, it has been observed in children aged as young as eight years. Red-cell agglutinins make their appearance later, but with the same frequency as in normal controls. There is insufficient data to pinpoint the time of appearance of rheumatoid factor and cryoglobulins.

What is the nature of this disturbance of IgM metabolism on which attention has justifiably been focused? Recent work has shown that increased production of IgM is responsible for the macroglobulinaemia.(4) Yet in response to non-malarial antigens such as flagellin, specific IgM antibody production in the primary response is depressed, whereas secondary IgG responses are normal. Conventional tests of cell-mediated immunity are also normal. Does the abnormality lie in the immune response to malaria, and to that alone? Patent parasitaemias are uncommon in established TSS, other evidence of significant parasitaemia, such as malaria pigment in macrophages or circulating monocytes, is also rarely seen, yet prolonged administration of antimalarials causes slow regression of all features of the disease, and rapid recurrence follows cessation

of therapy. All these observations indicate a prolonged and
excessive response to limited antigenic stimulation. It should be
emphasized, however, that IgG IFA titres are also significantly
elevated in TSS; and that IgG production is also significantly
higher than in controls, though its effect on serum IgG
concentrations is largely masked by the concomitant elevation in
plasma volume. Thus the defect may well relate to both IgM and
IgG responses to certain malarial antigens. The tribal and
familial distribution of the syndrome, both in Uganda and Papua
New Guinea, suggests a genetic basis, which may be related to the
inability to produce specific high avidity antibody capable of
effecting rapid clearance of antigen. If only low-avidity antibody
is synthesized, then both excessive production and persistence of
these antibodies in the circulation for prolonged periods, probably
complexed to malarial antigens, may result.

The cryoproteins which can frequently be found in the serum in
TSS contain IgG, IgM and complement, some of the antibody possessing
anti-malarial activity. No antigen has yet been identified, but it
can now be accepted that these cryoproteins represent immune
complexes of malarial antigen and antibody. The role of low
molecular weight malarial immune complexes in the production of both
transient acute nephritis and the quartan nephrotic syndrome is
established; and a comparable role for the cryoproteins of TSS has
been suggested in its pathogenesis.(5) The characteristics of
inclusions in the Kupffer cells indicate that ingestion of cryo-
immune complexes by these cells occurs, and may give rise to HSL
by mechanisms suggested by Greenwood.(6) Felty's syndrome may
provide an appropriate analogy.

Similar cryoproteins occur, though less frequently, in control
subjects; and one cannot at present state whether those of TSS are
merely present in higher concentration as reported by Greenwood,(6)
or whether they differ in their composition. Their demonstration
in splenic macrophages indicates that they may well contribute to
the splenomegaly directly by stressing the splenic phagocytic
mechanisms, and here again the appropriate analogy may be with
Felty's syndrome.

For many patients anaemia is the most debilitating facet of the
syndrome. Chronic depression of haemoglobin levels is almost
invariable and does not demand an immunological explanation. All
three mechanisms responsible - expanded plasma volume, shortened
red-cell survival and splenic red-cell pool - may occur in
splenomegaly of any cause; and it has been suggested that the
haematological abnormalities in TSS are no greater than those in

other diseases producing splenomegaly of the same degree. Direct Coombs' tests in many competent hands have been negative for both immunoglobulin and complement, and the two reports to date of contrary findings - one of positive direct Coombs' tests in two patients, using anticomplement antiserum, and the other of red cell autoantibodies detectable only by an enzyme-Coombs' test at 4° in 33% of subjects with TSS (the same frequency as in non-malarious diseases) - cannot be accepted without reservation as indicating autoimmune haemolytic anaemia as a constant and characteristic finding.

In the acute haemolytic crises of TSS, on the other hand, the association with pregnancy and the history of a premonitory febrile illness suggest a malarial trigger. The intense erythrophagocytosis in the spleen, and the occasional response to corticosteroids, suggest antibody mediation. Persistent immune complexes, formed during a malarial parasitaemia, and activating complement which then mediates red-cell destruction, could provide the explanation for this phenomenon which has not yet been adequately investigated.

Finally, what conclusions can be drawn about the pathogenesis of the disease from the effects of therapy? Historically, splenectomy preceded prolonged antimalarial therapy, and rapid correction of all the haematological abnormalities was demonstrated from various centres. IgG and IgM levels often fell rapidly in the early postoperative phase, but subsequently they slowly returned to preoperative values. Splenectomy did not significantly alter the impaired primary IgM response to non-malarial antigens, and it has been impossible to define any significant effect on the susceptibility of patients to infections.(7) These findings would indicate that the splenomegaly has profound influences on haematological parameters, but that the major disease manifestations do not result from the enlarged spleen as such and are influenced to a negligible degree by its removal.

Antimalarial therapy, on the other hand, has been shown in individual cases to reverse all manifestations of the syndrome; and this once again incriminates the immune response to malaria as the fundamental abnormality. Splenomegaly thus arises in response to stresses produced by this disturbance - it is a secondary phenomenon. Phagocytosis of parasitized or immunologically damaged red cells; dilatation of sinuses and extrasinusoidal spaces to create the splenic red cell pool; phagocytosis of or perhaps macrophage blockade by cryoprotein immune complexes; and lymphoid cell proliferation in response to the demand for antibody - all contribute. And all, exceptionally slowly, can be blocked by elimination of the appropriate antigen. But there are still a number of key points in the sequence that require confirmation before this hypothesis can be considered established.

SUMMARY

Studies of the pathogenesis of tropical splenomegaly syndrome (TSS) have generally failed to reveal an abnormal pattern of malarial infection, or any additional etiological agent. Field studies have shown that immunity to malaria, as judged by parasite rates at different ages, develops normally. Serum IgM levels, however, are disproportionately elevated even in young children prior to the acquisition of any other hallmarks of TSS, and continue to rise markedly throughout life. The IgM consists largely of immunofluorescent antimalarial antibody (IFA) and is a major constituent of the cold-precipitable immune complexes which appear responsible for many of the major manifestations of TSS. The role of these complexes in producing anaemia, the most important clinical finding, remains contentious.

The infrequency with which circulating parasites can be found, the slow resolution of the disease in subjects protected against malaria, and its rapid reappearance on re-exposure to infection, indicate a prolonged and excessive response to malarial antigens. The presence of high IgG IFA titres, and qualitatively normal responses to challenge with other antigens, exclude a general failure of the IgM/IgG switch as the cause. Conventional tests of T lymphocyte function are normal. There may, however, be a genetically-determined disturbance of antimalarial antibody production, resulting in the production of low-avidity antibody which complexes with malarial antigens rather than effecting their rapid removal.

ACKNOWLEDGEMENTS

The assistance of the Papua New Guinea Department of Public Health, and the financial support of the Wellcome Trust, in performing the studies described here, are gratefully acknowledged.

REFERENCES

1. MARSDEN, P.D. and CRANE, G.G. The tropical splenomegaly syndrome. A current appraisal. Revista do Instituto de Medicina Tropical de Sao Paulo, 18: 54 (1976).

2. CRANE, G.G. The pathogenesis of tropical splenomegaly syndrome – the role of immune complexes. Papua New Guinea Medical Journal, 20: 6 (1977).

3. CRANE, G.G. ET AL. Malarial antibodies in tropical splenomegaly syndrome in Papua New Guinea. Transactions of the Royal Society of Tropical Medicine and Hygiene, 71: 308 (1977).

4. FAKUNLE, Y.M. and GREENWOOD, B.M. Metabolism of IgM in tropical splenomegaly syndrome. Transactions of the Royal Society of Tropical Medicine and Hygiene, 70: 346 (1976).

5. ZIEGLER, J.L. Cryoglobulinaemia in tropical splenomegaly syndrome. Clinical and Experimental Immunology, 15: 65 (1973).

6. GREENWOOD, B.M. Proceedings of this meeting.

7. CRANE, G.G. ET AL. Tropical splenomegaly syndrome in New Guinea. II. Long term results of splenectomy. Transactions of the Royal Society of Tropical Medicine and Hygiene, 66: 733 (1972).

DISCUSSION

DAVID-WEST: At Ibadan, hepatic sinusoidal lymphocytosis is not seen in 1/3 of the patients and the grading varies in different parts of the liver when we looked at post mortem cases. The HSL does not correlate with spleen size, peripheral blood count or serum IgM. So I often wonder how important this finding is to the diagnosis or pathogenesis of the syndrome. It is however very interesting that Dr Fakunle has found immune complexes in the liver.

WEISS: This appears to be a haemolytic anaemia dependent upon a very large spleen. What is the reticulocytic count in this syndrome? What is their red-blood cell life span, and the life span of normal red-blood cells in these patients?

CRANE: I agree. In the steady state, bone marrow is hyperactive, reticulocyte counts are elevated, and red-cell ^{51}Cr T$\frac{1}{2}$ is shortened, typically to 10-14 days. The survival of donor red cells is much the same. In the acute haemolytic episode ^{51}Cr T$\frac{1}{2}$ may be as short as 2-3 days, with very high reticulocyte counts (up to 50%) and the appearance in the circulation in most instances of nucleated red cells.

HUMPHREY: Can the portal hypertension be explained by a presumably greatly increased splenic blood flow?

Can the anaemia be explained simply in terms of greatly increased vascular bed and quantity of macrophages in the spleen?

It has been reported that in juvenile rheumatoid arthritis autoantibodies against lymphocytes are present which appear to be selective for a suppressor cell sub-class (Strelkauskas - Schlossman - to be published in the Journal of Immunology). Would it be worthwhile looking for such anti-lymphocyte antibodies in TSS?

CRANE: Yes, in most cases (splenic blood flow can be up to 25% of the cardiac output). However, in some patients with complicating hepatic fibrosis there appears to be an obstruction to portal outflow as well.

The major factor responsible for the chronic anaemia is an expanded plasma volume. In part this results from the immunoglobuli overproduction, but not completely, and an expanded vascular space may also play a part. Certainly the haemolytic element may be related to the macrophage situation in the red pulp of the spleen.

There are a number of parallels with rheumatoid and particularly with Felty's – the cryoproteins, the hepatic sinusoidal lymphocytosis and the failure to mount an effective neutrophil response to bacterial infections. It would therefore be well worth explaining other similarities such as this one.

DAVID-WEST: In your slide showing malaria antibody titres in various age groups, are the titres related to serum IgM levels of the patients in each age group? To which malaria antigen is the antibody directed?

CRANE: There is close correlation between known IgM levels and IgM IFA titres in all subjects, and this is true of titres to Plasmodium falciparum, cynomolgi and brasilianum.

COHEN: Is it established that the high degree of resistance to malaria in TSS is directed against erythrocytic forms of the parasite? Is it possible that the liver pathology is associated with failure of exo-erythrocytic forms to develop after sporozoite infection?

GREENWOOD: As far as I know, no-one has looked at anti-sporozoite immunity in patients with TSS.

McGREGOR: In the Gambia where P. falciparum malaria is hyperendemic, a sizeable proportion of the adult population of rural villages develop serum levels of IgM similar to those found by Dr Crane in his TSS cases. However, TSS is rarely found in these persons. When sera from such populations are tested against batteries of soluble antigens recovered from the sera of children experiencing severe P. falciparum malaria we find that the persons identified as possessing antibodies to these antigens are also the individuals with high serum IgM levels. Wilson's studies on the antigens associated with erythrocytic forms of P. falciparum have indicated that the serum antigens from malarious children are heat-stable members of the S-class and that such antigens appear to be components of the host erythrocyte that have been modified by the parasite. Has Dr Crane any knowledge of the frequency of antibodies to this class of antigen in the sera of his TSS cases?

CRANE: This situation almost constitutes TSS without the splenomegaly, on the face of it, and it would be very interesting to establish whether these patients also have cryoprecipitable immune complexes in their serum – I suspect they will not, and that this will explain their failure to develop from splenomegaly. This also raises the question of some qualitative defect in the antimalarial IgM produced in TSS. Iain Wilson studied some of our TSS and control New Guinean sera some years ago. The controls had much the same frequency of antibodies to S antigens as did Gambian subjects, whereas there was a higher frequency in adults with TSS. In this respect TSS resembled your subjects with new high IgM levels, but I don't know exactly how comparable the findings are.

OZERETSKOVSKAYA: Is it possible that the prolonged antimalarial therapy reduced the TSS not by the antimalarial activity of the drugs, but by their pharmacological desensitization action?

CRANE: Our field studies showed that proquanil, chloraquine and cycloguanil palmoate gave comparable clinical responses. Therefore I feel that the effects must have been due to the abolition of parasitaemias, since the other pharmacological effects are very different.

OZERETSKOVSKAYA: What is the difference between the effect of enlarged spleen in TSS and in the other parasitic diseases with the enlarged spleen, for instance leishmaniasis or hydatid disease, and alveococcosis? In both cases the symptomatology of hypersplenism appeared?

CRANE: Apart from the acute haemolytic episodes, the haematological consequences are much the same, for a comparable degree of splenomegaly. However, the expansion of plasma volume and the consequent dilutional anaemia tend to be more marked in TSS than in some of the diseases you mention, because of the intense overproduction of immunoglobulins which substantially constitute to this hypervolaemia.

CRANE: Once you eliminate circulating parasites with antimalarials, it takes an extraordinarily long time (months to years) to produce complete reversal of the syndrome. Is it possible that immune complexes can persist for this long? Also, could a genetically-determined inability to produce high avidity antimalarial antibody produce disturbances of the magnitude we see in TSS?

ALLISON: Some antigens persist for long periods and the same can be true of immune complexes associated with dendritic cells. In relation to antibody affinity, one well-studied form of immune regulation is the feedback inhibition by IgG antibodies of IgM antibody formation. It may well be the case that low-affinity IgG antibodies cannot exert this type of feedback inhibition efficiently. The same may be true of anti-idiotypic antibodies if these are important controlling elements as the network theory proposes.

FRANK: I am struck by the fact that these patients have so few manifestations of immune-complex disease in the face of such high levels of immune complexes. I take it that they do not develop glomerulonephritis vasculitis etc.

Later I will discuss some aspects of the questions of what antibody and complement do to red-cell survival. I cannot help but note certain parallels. A very limited number of IgM molecules can markedly shorten the survival of red cells. The mechanism is complement dependent and involves the very rapid activation of complement on the erythrocyte surface. Under ordinary circumstances the red cell is not destroyed efficiently but if the animal has had an intercurrent infection with activated macrophages, we have studied BCG-infected animals, the destruction becomes very efficient and the cells are destroyed.

BATTISTO: The phenomenon you have described appears to be an IgM to IgG switch that has got stuck. The stimulation of the IgM could be attributable to a T-cell independent antigen. Have any attempts been made to stimulate T-cell replacing factor early in the disease as for instance by injecting KLH to try to get the pathway to IgG initiated?

CRANE: I think the balance of evidence is against the IgM-IgG switch being stuck in terms of the humoral immune responses generally. In response to flagellin primary IgM response was somewhat impaired, but secondary IgG response is normal in TSS. They have increased total IgG levels and increased IgG antimalarial IFA, and normal immunity to malaria, suggesting normal production of IgG inhibiting antibody. So if your suggestion were correct, it would have to be a specific defect relating to one or a limited range of antigens. In reply to the second part of the question, we haven't seriously considered doing this, and I wonder what one could reasonably use in man.

MORRIS: What does splenectomy do to TSS? Do they need continuing antimalarial treatment? What happened to your patients who had a splenectomy for the haematological disorder before antimalarial drugs were used? Finally, in an ideal world, would the appropriate management of the patient with severe TSS be splenectomy and antimalarial treatment?

CRANE: Splenectomy corrects the haematological abnormalities, but in the long run does not alter the other manifestations of the syndrome. In our patients morbidity was strikingly reduced, and mortality over 4-6 years of follow up was significantly less than in

untreated patients. Deaths were usually due to overwhelming acute
febrile illnesses, as is also the case in the untreated disease.
And as these usually occurred in villages, one could not determine
whether they were bacterial or malarial in origin. Equally, one
cannot clearly attribute them to the consequences of splenectomy
rather than to the basic impaired resistance to infection. Long-
term antimalarials are almost always effective, and I doubt if
splenectomy would even now be justified as a prelude to definitive
drug therapy - even though clinical improvement with drugs may occur
very slowly.

CLAYTON: I have been wondering about the cells making all this
immunoglobulin. Either there are very long-lived cells
continuously secreting it, which from most evidence seems unlikely,
or else there is a constant supply of new cells which mature. Can
you tell me whether all these cells are coming from?

GREENWOOD: The bone marrow appears very active suggesting a high
turn-over of lymphocytes.

CLAYTON: Does all this intense proliferation every lead on to
transformation into a true leukaemia, as is thought to happen with
other neoplasias?

DAVID-WEST: In Nigeria we had difficulties in distinguishing cases
of chronic lymphatic leukaemia often presenting without splenomegaly
from cases of TSS presenting with leucocytosis and a lymphocytic
leukaemoid reaction. Serum IgM and PHA transformation of lymphocytes
can now differentiate between the two conditions. One wonders,
however, whether there is a continuous spectrum in these conditions
especially taking into consideration the case just described of
malignant change in a case of TSS. Again longitudinal studies and
good initial documentation of cases is needed.

ALLISON: I believe that you have some evidence that what appears
to be a typical TSS can later be manifested as a lympho-reticular
malignancy. Would you care to comment on this?

DAVID-WEST: We have been fortunate at Ibadan to be able to follow
up the majority of our TSS patients for many years, over 10 years
in many cases, while they are on antimalarial therapy. One patient
whose spleen was originally over 20 cm was treated for a year with
proquanil and the spleen regressed. She defaulted treatment for
18 months and when seen her spleen was larger than on first
presentation and a lymph-node biopsy showed a "lymphomatous reaction"
She did not respond to antimalarial therapy this time and died

shortly after. Post mortem confirmed a diagnosis of malignant lymphoma. We have thus wondered whether the prolonged reticuloendothelial stimulation could progress to a malignant proliferation if not arrested in time. Long-term follow up of cases of TSS is needed and I wonder whether other centres have a similar experience.

POLTERA: Did anybody try to isolate antigen(s) from circulating immune complexes in TSS, and if so have such isolated antigens been exposed to specific malarial antibodies?

CRANE: Alister Voller's group have tried to identify malarial antigens in cryoproteins by countercurrent electrophoretic techniques before and after standard manoeuvres for disassociating the components of such complexes, with no success. Therefore we have done no further work on the antigens. The complexes do contain malarial antibody activity, as detected by indirect immunofluorescence.

MITCHELL: Brian, were you surprised that the half-life of injected IgM was not reduced in macroglobulinaemias?

GREENWOOD: No. This has not been found to be the case in other forms of macroglobulinaemia in contrast to the situation in subjects with elevated serum IgG levels.

ALLISON: Dr Fakunle, would you care to amplify any of Dr Greenwood's statements?

FAKUNLE: Dr Greenwood has already emphasized that the data on characteristics of the HSL as predominently T cells and their sensitization to a crude liver extract are preliminary. Techniques are now available to determine lymphocyte subtype in tissue sections and the migration inhibition experiments need to be repeated with various types of liver specific lipoproteins. It should, however, be emphasized that in addition to HSL, infiltration of portal tracts with mononuclear cells and mild portal fibrosis occurs in a considerable number of patients with TSS. If these lymphocytes are damaging the liver cells the resultant fibrosis could contribute to the portal hypertension in these patients in addition to the increased portal blood flow. We hope to carry out further studies along these lines.

SPRENT: Recent studies of Coutinho et al. in Basel might have a possible bearing on the etiology of TSS. These workers raised anti-idiotypic antisera to murine myeloma proteins and made an unexpected

258

discovery that their antisera stained a high proportion (15-20%) of normal B cells of any mouse strain tested. Moreover the antisera acted as a polyclonal mitogen for B cells. From this and other evidence the authors draw the provocative conclusion that "idiotypic" determinants are present not only on immunoglobulin receptors but also on mitogen receptors, each B cell having a spectrum of different mitogen receptors. So perhaps one might consider the outside possibility that TSS is in part a manifestation of chronic polyclonal activation of B cells stimulated by anti-idiotypic antibodies formed against antimalarial antibody. In this respect it would be interesting to know whether serum from patients with TSS would act as a polyclonal activation for B cells of normal patients.

The Role of the Spleen in the Origin of Some Clinical and Immunological Features of Alveococcosis and Echinococcosis

N.N. Ozeretskovskaya, N.I. Tumolskaya, A.M. Poverenny,
V.V. Serov, V.K. Podgorodnitchenko, A.E. Sivakov,
E.I. Martsinovsky Institute of Medical Parasitology and Tropical
Medicine, Institute of Medical Radiology of the Academy of
Medical Sciences, Chair of Pathologic Anatomy of the 1st
Moscow Medical Institute, Moscow, USSR.

Echinococcosis (Echinococcus granulosus infestation) and
alveococcosis (Echinococcus multilocularis infection) are two of
the most severe larval helminthiases of humans with progessive
course of disease and high mortality.   Enlargement of the spleen
is detected in about half of the patients hospitalized, and seems
to be beneficial in respect of the severity of disease.   In 1974
in the USSR and in the USA, independently, publications
appeared on the high chemotherapeutic efficacy of mebendazole
(carbamate-benzimidazole derivative) in alveococcosis.   While
effective under experimental laboratory conditions, mebendazole did
not show sufficient curative action in the advanced, severe form of
the disease.   It seemed to hinder initially the progressive course
of the disease but became ineffective subsequently, in spite of the
prolonged treatment and elevated doses.

The role of the spleen in the clinical and immunological
conditions in echinococcosis and alveococcosis, as models of severe
chronic parasitic diseases, was studied and, especially, the
influence of the enlarged spleen upon the chemotherapeutic and
tide effects of mebendazole was analyzed.

Thirty-five patients, 22 suffering from alveococcosis and 13
with echinococcosis, including 15 males and 20 females (females
predominated in the alveococcosis group)(Table I), were examined
under clinical conditions for between three and 17 years.   Seventeen

TABLE I

CLINICAL DATA ON ECHINOCOCCOSIS AND ALVEOCOCCOSIS PATIENTS TREATED WITH MEBENDAZOLE

Diagnosis	Number of patients	Males	Females	Symptomatic period of disease	Patients with the spleen		Complications			Surgical treatment		Chemotherapy (number of courses)
					normal	enlarged	Necrosis, disintegration, fistula	Metastases	Jaundice	1-2 times	Repeated	
Alveococcosis	22	8	14	11.65+ +1.59	9	13	9	4	4	14	4	6.19+ +0.5
Echinococcosis	13	7	6	16.58+ +2.8	8	5	3	4	2	5	9	5.58+ +0.83
Total	35	15	20		17	18	12	8	6	19	13	

out of 35 patients had spleens of normal zize, in 16 the spleen was enlarged. Most of the patients had severe and extra-severe disease. There was a distinct preponderance of the patients with a non-enlarged spleen in the latter group (Table II).

In 12 cases the disease was complicated by liver decomposition, necrosis and fistulae or collapsed cysts; eight patients had metastastic damage of the lungs, bones or diseminated cysts in the abdominal cavity, five patients suffered from relapsing or continuous jaundice.

On the average, the symptomatic period of disease was longer in the echinococcosis patients than in those with alveococcosis. Both groups of patients had been treated surgically. Only three of the 22 alveococcosis patients had had explorative operations; 14 patients had been operated on once or twice, the other five had been operated on repeatedly. All the 13 echinococcosis patients had been surgically treated, four of them once or twice and nine repeatedly. The number of courses of chemotherapy in both groups was equal (Table I).

Routine clinical, laboratory and roentgeno-radiological examinations were performed. The levels of specific antibodies, using Echinococcus granulosus crude antigen in the passive haemagglutination test (PHAT) were estimated, determination of the serum immunoglobulins using the Tri-Partigen immunodiffusion plates was performed, levels of antibodies to the native DNA (nDNA) and DNA, denatured by heating (dDNA), by immunoradiological technique, using.the absorption of immune complexes upon the nitrocellulose membranes (1) and the levels of antibodies to DNA, completely denatured by heating in formaldehyde solution (fDNA) by the PHAT (2) were studied. The phytohaemagglutinin (PHA) skin test, indicative of the T lymphocytes potency (in vivo blast transformation test, (3)) was also performed.

It is well known that the main pathology in multiple echinococcosis and in alveococcosis is damage of the liver and kidneys. In spite of the predominant localization of the parasite's larvae in the liver, especially in alveococcosis, and the rarity of direct damage to the kidneys, renal failure is one of the most frequent causes of death.

TABLE II

SEVERITY OF DISEASE IN ALVEOCOCCOSIS AND ECHINOCOCCOSIS PATIENTS
WITH NORMAL AND ENLARGED SPLEENS

Diagnosis	Total number of patients	Size of the spleen	Clinical course of disease		
			Moderate	Severe	Extra severe
Alveococcosis	22	Normal	2	7 ⟍12	5
		Enlarged	2	4 ⟍6	2
Echinococcosis	13	Normal	3	5 ⟍6	1
		Enlarged	3	1 ⟍1	-
Total	35		10	17	8

The signs of renal disorders, such as proteinuria, haematuria, the appearance of the hyaline casts, were established in 25 out of 35 patients. In seven some features of more profound damage to the kidney, such as isosthenurua, and impaired filtrative and reabsorption functions appeared.

A significant preponderance of persons with non-enlarged spleen among those with kidney damage was found (Table III). The serum IgG

TABLE III

RENAL DAMAGE IN ECHINOCOCCOSIS AND ALVEOCOCCOSIS PATIENTS
WITH NORMAL AND ENLARGED SPLEENS

Renal damage	Not revealed	Moderate		Severe	Total
Patients with normal size of spleen	3	9	17	8	20
Patients with enlarged spleen	7	3	8	5	15
Total	10	12		13	35

$\chi = 4.1975$ (p < 0.05)

level in this group was lower and there was also a tendency to elevation of the amount of IgM (Table IV). These changes in the serum immunoglobulin levels correlated with the significantly higher level of the total serum protein and percentage of γ-globulins in patients with splenomegaly (Table V). Also, levels of specific antibodies in the PHA test in patients with enlarged spleen, were significantly higher than in those with normal sized organ (1:22727 and 1:640 respectively). After the courses of treatment the specific antibody levels decreased but the difference between the both groups of patients remained 1:5650 and 1:285 respectively).

In alveococcosis patients the PHA skin test produced some difference between persons with normal and enlarged spleen (12.5+3.12 and 18.33+3.34), but in echinococcosis the tendency was

TABLE IV

SERUM IMMUNOGLOBULIN LEVELS IN ECHINOCOCCOSIS AND
ALVEOCOCCOSIS PATIENTS WITH NORMAL AND ENLARGED SPLEENS

Size of spleen	IgA	IgG	IgM
Normal	273.33 ± 25.98	1765.0 ± 103.37	236.3 ± 32.15
Enlarged	283.08 ± 40.78	2906.43 ± 123.18	198.0 ± 19.06
p	> 0.05	< 0.01	> 0.05

reversed (25.0 ± 3.96 and 15.8 ± 2.4, $t = 1.98$). On average there was
a significant difference between the nDNA binding capacity of the
serum between patients with enlarged and normal spleen (Table VI).
If scrutinized separately, the difference between the
echinococcosis and alveococcosis patients with normal and enlarged
spleens was insignificant, perhaps due to the small number of cases.

Kidney biopsy in patients with nephrological disorders and
examination of bioptates by histological and immunofluorescent
(indirect immunofluorescence test) techniques demonstrated
proliferation of the glomerular cells, periglomerular sclerosis,
tubular dystrophy and glomerular deposits of IgG.

Of the most interest was the changes in immunological features
during prolonged courses of the treatment with mebendazole. The
level of PHA-skin test in patients with the unenlarged spleen was
initially quite normal and in the splenomegalic elevated once or
twice (Fig. 1). During repeated courses of treatment, the skin
reaction gradually decreased but with the same difference of levels.
At the time of the 5-6th course, the size of papula in the
patients of the first group dropped to the minimal normal level and
with the 11-12th courses became negative. In the second group the
minimal normal level of the PHA test remained until the 11th course
(Fig. 1).

The same changes were estimated in fDNA-antibody levels (Fig. 2)
which showed normal data at the time of the 9-10th course of
medication in the first group of patients and remained significantly
enlarged in the second group until the 13-14th courses (Fig. 2).

TABLE V

TOTAL SERUM PROTEIN CONTENTS AND PROTEINOGRAMME IN ECHINOCOCCOSIS AND ALVEOCOCCOSIS PATIENTS WITH NORMAL AND ENLARGED SPLEENS

Size of spleen	Total protein (g/per cent)	Proteinogramme (per cent)					
		Albumins	Globulins				
			α_1	α_2	β	γ	
Normal	7.71 ± 0.12	51.55 ± 3.21	5.34 ± 0.73	7.8 ± 0.6	13.41 ± 0.83	21.82 ± 1.93	
Enlarged	8.82 ± 0.14	38.27 ± 1.15	5.24 ± 0.57	8.44 ± 0.7	13.43 ± 1.17	34.51 ± 1.78	
p	< 0.001	< 0.01	> 0.05	> 0.05	> 0.05	< 0.01	

TABLE VI

SERUM nDNA AND dDNA BINDING CAPACITY IN ECHINOCOCCOSIS AND ALVEOCOCCOSIS PATIENTS WITH NORMAL AND ENLARGED SPLEENS

| Size of spleen | nDNA | | dDNA |
	in 0.15M NaCl solution	in 0.05M NaCl solution	(in 0.15M NaCl) solution
Normal	10.06 $\pm$ 1.25	23.3 $\pm$ 3.3	9.09 $\pm$ 0.98
Enlarged	17.2 $\pm$ 2.02	24.95 $\pm$ 3.74	9.62 $\pm$ 0.98
p	< 0.01	> 0.05	> 0.05

The most interesting data were obtained by examination of the changes of nDNA serum-binding capacity, indicative of the immunopathological tendency of the disease. Almost equal initial levels of the nDNA antibodies were found in both groups of patients, with some tendency towards lower levels in the second group (Fig. 3) During the courses of treatment the amount of nDNA-antibodies in the first group of patients decreased to normal, whereas in the second group, after the primary normalization at the time of the 3-6th courses, it began to rise dramatically, exceeding the initial level (Fig. 3) by the time of 11-12th courses.

The data obtained permit some preliminary conclusions. It seem that the reactivity of the spleen plays an important role in the clinical features and the course of echinococcosis and alveococcosis Patients with splenic activity benefit in respect of diminuation of the severity of disease and especially renal damage. Enlargement of the spleen correlates with the higher (normal) level of the skin PHA-test, higher production of serum IgG and a smaller amount of IgM The glomerular deposits of immune complexes, containing IgG immunoglobulins, in patients with renal disorders permits the supposition of immune complex mediated kidney damage and, perhaps, a protective role for the enlarged spleen in the development of immune-complex disease.

Prolonged mebendazole treatment, producing the degradation of parasite's larvae, controls massive antigen entrance. It seems that in these conditions, the enlarged (or fibrosed) spleen at some

FIGURE 1

PHA-SKIN TEST IN ECHINOCOCCOSIS AND ALVEOCOCCOSIS PATIENTS
WITH NORMAL (o) AND ENLARGED (●) SPLEENS
DURING PROLONGED MEBENDAZOLE TREATMENT

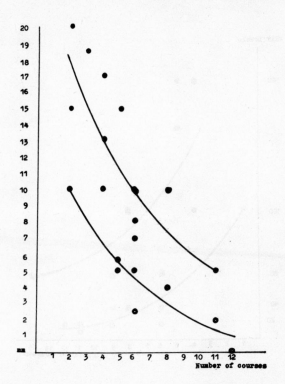

critical point may be the cause of the elevated autoantibody
production. In every case, prolonged mebendazole treatment in
patients with the advanced echinococcosis and alveococcosis seems to
be one of the factors stimulating immunological abnormalities.

268

FIGURE 2

ANTIBODIES TO fDNA IN THE PHAT IN ECHINOCOCCOSIS AND
ALVEOCOCCOSIS PATIENTS WITH NORMAL (o) AND ENLARGED (•)
SPLEENS DURING PROLONGED MEBENDAZOLE TREATMENT

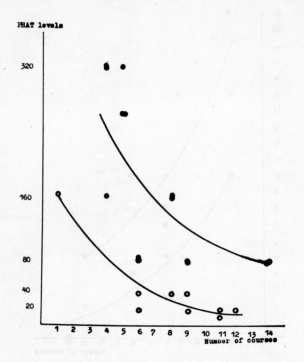

FIGURE 3

SERUM nDNA BINDING CAPACITY IN ECHINOCOCCOSIS AND
ALVEOCOCCOSIS PATIENTS WITH NORMAL (o) AND ENLARGED (•)
SPLEENS DURING THE PROLONGED MEBENDAZOLE TREATMENT

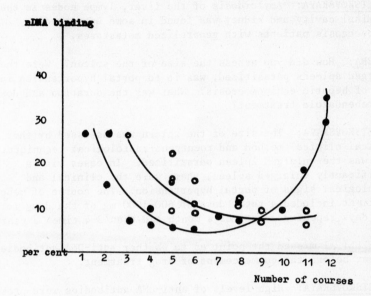

REFERENCES

1. PINCUS, T. Immunochemical conditions affecting the measurement
   of DNA antibodies using ammonium sulfate precipitation.
   Arthritis and Rheumatism, 14: 623 (1971).

2. POVERENNY, A.M. and LEVI, M.I. The existence of two types of
   antibody to DNA. Voprosi Mediciniscoi Chimii, II: 2.95 (1965).

3. BLAESE, R.M. ET AL. Phytohemagglutinin as a skin test for the
   evaluation of cellular immune competence in man. Journal of
   Laboratory and Clinical Medicine, 81: 538 (1973).

DISCUSSION

MITCHELL: In what proportion of treated or untreated patients is amyloid detected in these larval cestode infections and what organs are involved?

OZERETSKOVSKAYA: Amyloidosis of the liver, lymph nodes in the abdominal cavity and kidney was found in some untreated alveococcosis patients with generalized metastases.

POLTERA: How did you assess the size of the spleen? Were the enlarged spleens parasitized, was it to portal hypertension in the case of hepatic echinococcosis? What was the duration and dose of the mebendazole treatment?

OZERETSKOVSKAYA: The size of the spleen was assessed by the physical clinical method and roentgeno-radiological technique. In case was the enlarged spleen parasitized. In cases with a significantly enlarged spleen, there were the clinical and radiological signs of portal hypertension. The course of mebendazole treatment included a total dose of 600-1000 mg of the drug for 5-10 days repeated 3-4 times a month and then 3-4 times a year.

MITCHELL: I missed the point as to whether anti-DNA antibodies were present before the commencement of drug treatment.

OZERETSKOVSKAYA: High levels of anti-DNA antibodies were present before the beginning of chemotherapy. For example, up to 30-50, and in some cases 20% of nDNA-binding capacity of serum in 0.15M NaCl solution and up to 30-50% in 0.05M NaCL solution which in the case of nDNA binding capacity in high ionic strength solution (0.15M NaCl) was comparable with those in disseminated lupus erythematosus. During the first 3-5 courses the binding capacity of the serum decreased and then began to rise dramatically exceeding the initial level at the 11-12th courses. In our recent investigation on the meaning of the production of antibodies to DNA in the pathogenesis or immunopathology of parasitic diseases, we found an awfully high correlation in malaria between the production of fDNA antibodies and the severity of the disease. In cases of malignant subtertian malaria from South-East Asia, the levels of fDNA antibodies rose to 1:5000 - 1:10000 in PHA . In contrast, in giardians the slightly raised levels of nDNA-antibodies in 0.15M NaCl solution and significantly increased amounts of nDNA-antibodies in 0.05M NaCl solutions had no relation to the clinical features of the disease.

ALLISON: In the nephropathy associated with parasitic infections is DNA present in the renal glomerular complexes?

OZERETSKOVSKAYA: We haven't made the differentiation screening yet, the estimation is in progress.

The Role of the Reticuloendothelial System in the Clearance and Destruction of Circulating Cells Coated with Antibody and Complement

M.M. Frank, National Institute of Allergy and Infectious Diseases, National Institutes of Health, Bethesda, Maryland, USA.

During the course of this meeting it has been repeatedly emphasized that the reticuloendothelial system plays a key role in determining the course of many parasitic infections. We have discussed several diseases associated with rapid changes in the numbers of parasites in the circulation. We have heard of diseases like tropical splenomegaly in which splenomegaly may be associated with Coombs-positive haemolytic anaemia and is often associated with severe haemeolysis at the time of intercurrent infection. It has been repeatedly emphasized that the dissection of the various elements responsible for the pathophysiological manifestations is difficult in these diseases. The parasitic infestation is associated with the development of an immune response to multiple parasite antigens. The presence of the organism can lead to activation of lymphocytes and macrophages, and the development of suppressor lymphocytes and macrophages may be noted. The parasites may directly activate complement and other mediator systems in the presence and, at times, in the absence of antibody. All of these elements may influence the course of infection.

Although my own interest has not been in the interplay between parasites and reticuloendothelial elements, we have tried to simplify the experimental systems and to ask some direct questions about how antibody and complement on the surface of particulate materials in the circulation may influence the clearance of the

particle and the site in the reticuloendothelial system in which
the particle is cleared. Before we enter into a direct discussion
of these factors, let us consider a few of the elements of RES
function.

It has been recognized for years that the reticuloendothelial
system is of major importance in the clearance of foreign
substances from the blood stream.[1] Particular attention has been
paid to the clearance of bacteria and there is an extensive
literature on this subject. There is also a large body of
information available on the clearance of proteins which have been
chemically or physically modified and on the clearance of cells
which have been chemically or physically altered.[1,2]

Perhaps the area which has been explored in greatest detail
concerns the factors which influence the survival of erythrocytes
which have been damaged by immunological processes.[3-5] Early
workers interested in the pathophysiological basis of autoimmune
haemolytic anaemia made a series of important observations starting
in the late 1940's when ^{51}Cr became widely available for
erythrocyte survival studies and arrived at a series of general
conclusions. These can be expressed as follows: the liver is the
principal reticuloendothelial organ responsible for the removal of
erythrocytes which have sustained major damage to the cell surface.
The spleen, on the other hand, is responsible for the removal from
the circulation of cells which have sustained minor membrane damage
or alterations in shape-like spherocytes. In both cases it is
reticuloendothelial elements within these organs that are responsible
for the clearance function. In the case of the liver, the
reticuloendothelial elements responsible for the clearance are the
Kupffer cells lining hepatic sinusoids. In the spleen, macrophages
lining the sinusoids of the red pulp are the principal
reticuloendothelial elements. It was observed that when erythrocytes
were removed from men or animals, coated with anti-erythrocyte
antibody, and reinfused into the donor, one of a number of effects
of this coating was noted. If antibody and complement were found to
interact with the erythrocyte surface, the erythrocytes were cleared
by the liver. If, on the other hand, the antibody coating the
erythrocyte surface was of a type which did not activate the
complement system, the spleen was the site of clearance of the cells,
unless very large numbers of antibody molecules were bound to the
cell surface. If very large amounts of antibody were bound, cells
were cleared by the liver. A logical conclusion from this series
of observations was that the complement produced major membrane
damage and the cells were therefore cleared by the liver. On the
other hand, the perturbation to the cell membrane caused by antibody
which did not activate complement was relatively minor and such cells

were cleared by the spleen. With the discovery of a macrophage membrane receptor for the Fc portion of IgG, an additional layer of complexity was added to this general model and it was presumed that most clearance of IgG-coated cells was mediated via these Fc receptors.(6,7)

Over the last several years we have been involved in a careful re-exploration of these phenomena and have come to rather different conclusions.(8)[1] Although we agree that cells which have sustained major membrane damage are cleared by the liver and cells which have sustained more minor damage may often be cleared by the spleen, it has become evident that much of the clearance sustained by cells coated with immunologically-reactive materials, including complement protein fragments, is not due to non-specific damage to the cell membrane. It is rather due to the specific interaction of these immunologically-active proteins with receptors for these proteins located on the membrane of the cells of the reticuloendothelial system. Such receptors on reticuloendothelial macrophages have been demonstrated for the Fc portion of IgG as well as for a number of complement fragments. These include the opsonic fragment of C3, C3b, but also a cleavage fragment of C3b termed C3d, for a fragment of C4 termed C4b, and the C1q portion of C1. It is the interaction of the erythrocytes coated with immunologically-active proteins with these membrane receptors that determines their fate. Moreover, the nature of the coating determines where in the RES these cells are sequestered. Various coatings may cause preferential sequestration in the liver or the spleen.

Our first studies were of the effect of IgM antibody on erythrocyte survival.(9,10) We sensitized ^{51}Cr tagged guinea-pig erythrocytes with rabbit IgM antibody and followed the fate of these cells after reinjection. No receptors for the Fc fragment of rabbit IgM have been reported on guinea-pig macrophages. Nevertheless, when one coats erythrocytes with IgM antibody and injects them back into the guinea-pig, one finds that the cells are rapidly cleared from the circulation and are sequestered within the liver (Fig. 1). The rate of clearance is extremely rapid and sequestration occurs via adherence to the surface membrane if Kupffer cells within the sinusoids of the liver. The rate of sequestration in fact only appears to be limited by the rate of liver bloodflow. However, the majority of the cleared cells are not destroyed. After a period of

---

[1] Many students and colleagues participated in these studies and made them possible. These are named at the end of the text.

FIGURE 1

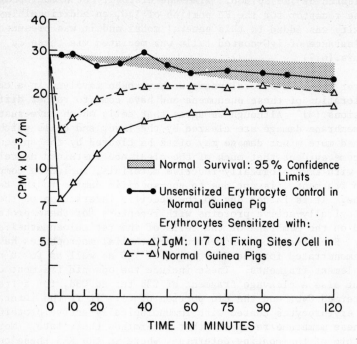

51Cr Clearance of IgM coated erythrocytes in normal guinea-pigs.
Cr labelled erythrocytes were coated with rabbit IgM antibody. An
average of 117 molecules per cell was used in the erythrocyte
coating, representing 117 C1 fixing sites per cell. The cells were
injected into normal guinea-pigs and showed rapid sequestration with
subsequent release into the circulation. (Reprinted from the
Journal of Clinical Investigation)

time the sequestered cells leave the hepatic micro-circulation and
return to the general circulation as "Coombs-positive" erythrocytes.
The circulating erythrocytes remain coated with antibody and
complement. Using C4 deficient and C3-9 depleted guinea-pigs we
showed that complement activation is an absolute requirement for
this clearance to take place and suggested that erythrocyte membrane
C3b was responsible for the clearance reaction. The data indicated
that when IgM-sensitized cells were introduced into the circulation
they rapidly activated the complement system. The rate of activation
was so rapid that, in fact, it played no role in the rate of

clearance of the cells which depended primarily upon the rate of liver bloodflow. In the guinea-pig model that we established it takes a minimum of 60 IgM molecules per erythrocyte to produce hepatic sequestration. Below this level of sensitization the cells had normal survival and at no level of sensitization were they cleared primarily by the spleen. In similar studies in man using IgM anti-A isoagglutinin, we showed that about 20 IgM molecules per cell will produce hepatic sequestration (Fig. 2).(11) The cells were cleared because complement, particularly the complement component fragment C3b, was deposited on their surface and this interacted with C3b receptors on the hepatic macrophages.

FIGURE 2

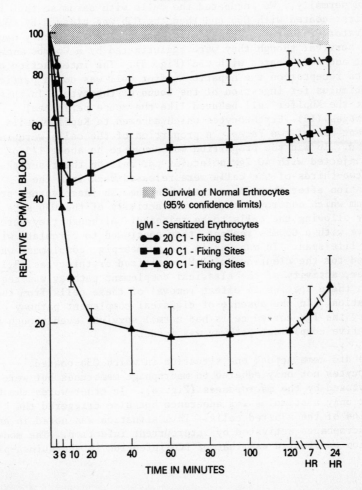

Clearance of IgM sensitized cells in man. As shown the cells were coated with 20, 40, and 80 molecules of IgM on an average per cell and injected intravenously. Again, hepatic sequestration was demonstrated. A portion of the cells, about two thirds, returned to the circulation thereafter to survive normally. (Reprinted from the Journal of Clinical Investigation)

It was possible to duplicate the in vivo events in vitro and to recreate in the test tube some of the processes which occurred during the clearance and release of the erythrocytes.(9,10,12) We incubated IgM-sensitized erythrocytes thereby coating them with C3b. When such cells were re-injected into animals they were cleared normally. We incubated the cells with serum so that they were first coated with C3b and then the C3b was cleaved by C3b inactivators in serum. These cells were not cleared; they had normal survival though they were agglutinated by a Coombs antiglobuli reagent and were coated with C3d (Fig. 3). The intereaction of C3b with its receptor on the hepatic Kupffer cell was not a particularly good stimulus for ingestion of the bound erythrocyte. In this respect the Kupffer cell behaved like the mouse peritoneal macrophages.(13) Erythrocytes which adhered to Kupffer cells underwent one of two fates. A proportion of the cells were indeed ingested. In man the proportion was shown to be about 1/3 of cells injected with 40 IgM molecules per cell on their surface. About two-thirds of the cells were released back into the circulation after they interacted with the C3b inactivators present in serum which destroyed the opsonic activity of the C3b site, thereby allowing the cells to be released. Although they were still positive with a Coombs reagent, they continued to circulate with a normal life span. In man, as in the guinea-pig, complement was required for the clearance of IgM sensitized erythrocytes.(9,11) Moreover, activity of the classical complement pathway was required in both these species to effect removal of these cells from the circulation. In the absence of classical complement pathway activity the IgM coated cells had normal survival even though the alternative complement pathway was intact.

We did come across one situation in which C3b-coated erythrocytes not only adhered to macrophage membranes but were also phagocytosed by the macrophages (Fig. 4). In other words the C3b was not only a stimulus for adherence but also triggered the ingestion of the adhered cells. This situation was noted in animals with macrophages activated by intercurrent infection. The model infection we studied was that of BCG infection in the guinea-pig but

FIGURE 3

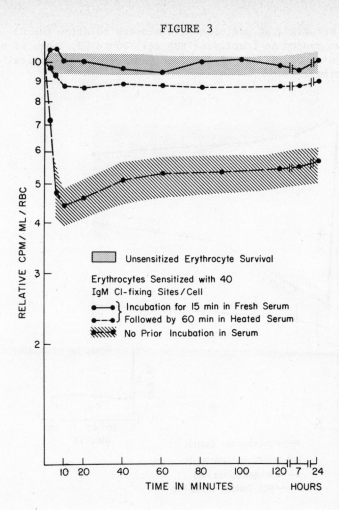

Clearance of cells coated with IgM antibody, 40 sites per cell, in man. The hatched area shows the clearance of cells not exposed to the patient's serum before injection. The line in the hatched area indicates cells exposed to fresh serum for 15 minutes before injection. The 15 minute exposure to fresh serum makes no difference in cell survival. However, when the cells are exposed to fresh serum for 15 minutes to allow deposition of C3b, and then

280

to heated serum for an additional 60 minutes to allow the C3
inactivator system to inactivate the cell bound C3, there is no
decrease in survival. (Reprinted from the Journal of Clinical
Investigation)

FIGURE 4

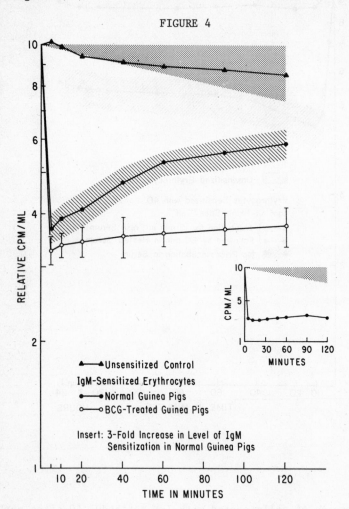

Clearance of IgM sensitized cells in normal guinea-pigs and
guinea-pigs infected with BCG. Clearance of cells with 234 C1
fixing sites per cell. In animals who had recovered from BCG

infection but with granuloma and activated macrophages persisting, initial clearance was slightly greater than in normal guinea-pigs. However, the cleared cells were not returned to the circulation and were phagocytosed by Kupffer cells within the liver. As shown by the insert, this situation could be duplicated by increasing the level of erythrocyte sensitization three-fold, thereby increasing the amount of C3 on the red cell surface greatly. (Reprinted from the Journal of Clinical Investigation)

the relevance to the material which has been presented concerning tropical splenomegaly should be clear. Cells coated with C3 which would not be sequestered and destroyed within the RES in the absence of infection were now both cleared and destroyed. Minimally sensitized cells could be damaged.

The effects of IgG antibodies were quite different (Fig. 3). (9,10) We confirmed earlier observations suggesting that the presence of IgG antibodies on the surface of isologous erythrocytes injected into experimental animals or man lead to progressive clearance of the erythrocytes from the circulation (Fig. 5).(3,4) This clearance was effected by the spleen. Unlike the IgM-coated cell, splenic sequestration was not followed by release of the sequestered cell back into the circulation but was followed promptly by phagocytosis of the sequestered cells and their destruction within the sinusoids of the spleen. In the guinea-pig, we demonstrated that complement plays a major role in the clearance of IgG-coated erythrocytes from the circulation.(14) We believe the complement may be quite important in the clearance of anti-Rh IgG-coated cells by the spleen in man although the sparse distribution of antigen on the erythrocyte surface precludes the formation of antibody doublets.(15,16) Nevertheless we have made the observation that no patient we have studied with a C4 level below about 1% of the normal level, has had normal IgG Rh antibody mediated clearance. Moreover, individuals with C2 deficiency also have abnormalities in their Rh-antibody-mediated clearance. We have suggested that small amounts of C3d may play a role in this Rh-antibody-mediated erythrocyte destruction. This has been shown to be true using other antibodies in in vitro studies;(17) however, the data in man on which this hypothesis rests is still incomplete.

There are two general areas for further investigation which grow out of these studies that I wish to comment on in some detail. The first concerns the application of these studies to other human diseases in which one is interested in knowing the status of the various receptor systems. Thus, it would be of great interest to know the status of the Fc and C3b receptors in patients with

FIGURE 5

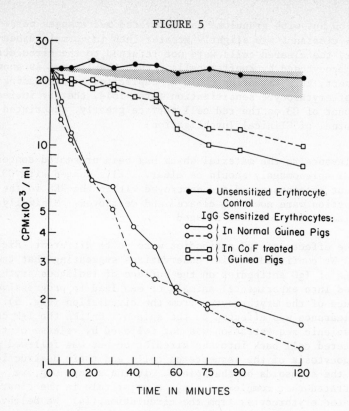

Clearance of IgG coated erythrocytes in guinea-pigs. Shown in the progressive clearance from the circulation of IgG coated erythrocytes with sequestration and phagocytosis in the spleen. The contribution of complement components underscored in that animals treated with cobra venum factor to deplete the late acting complement components, show a major decrease in the rate of sequestration. However, the site of sequestration remains the same. (Reprinted from the Journal of Clinical Investigation)

schistosomiasis, malaria, etc. In fact, we have shown that the techniques developed for the study of erythrocyte clearance in man can be used to access receptor function in a variety of diseases. Given the constraints in time, I will not develop these concepts

in detail but will simply list a series of observations which we
have made in some diseases which fall into the general category
of autoimmune diseases and present one set of observations in
greater detail.

We have examined two diseases of the liver which are thought
to have an aspect of immunological damage in their development.(18)
These are primary biliary cirrhosis and chronic active hepatitis.
In addition, we studied patients with large duct biliary obstruction
and alcoholic cirrhosis as controls. In brief we have found that
patients with primary biliary cirrhosis and chronic active hepatitis
as well as patients in the control groups had no defect in Fc
receptor function (Fig. 6). In fact these patients were able to

FIGURE 6

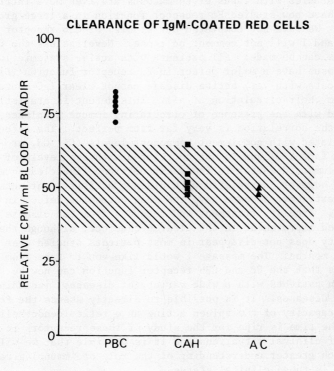

Clearance of IgM coated erythrocytes in patients with PBC,
primary biliary cirrhosis, CAH, chronic active hepatitis and AC,
alcoholic cirrhosis. Patients with primary biliary cirrhosis fail

to clear C3b coated erythrocytes from their circulation in a
normal fashion. This is not true of patients with the other chronic
liver diseases shown.

clear IgG-coated cells more rapidly from the circulation via splenic
sequestration than normal individuals. However, patients with
primary biliary cirrhosis but not patients with chronic active
hepatitis, alcoholic cirrhosis or large duct biliary obstruction
have a major defect in their C3b receptor activity and cannot clear
C3b coated erythrocytes normally from their circulation. This
defect is unrelated to the stage of disease. Interestingly, these
patients all have normal RES function as measured by the standard
test used which measures clearance of aggregated albumin.

Our studies with lupus erythematosus are even more intriguing.
(19) We have now studied Fc receptor function in a large group of
patients. We have not completed our studies of C3b receptor
function and I will not comment on those. Nevertheless the following
statements can be made: All patients with active systemic lupus
erythematosus have a major defect in Fc receptor function (Fig. 7).
Some patients with very active disease cannot clear IgG-coated
cells from their circulation at all. This defect is statistically
correlated with the presence of circulating immune complexes,
although the correlation is very far from perfect. The defect is
not correlated with complement component levels (C4, C3, or
properdin factor B) and is not correlated with the levels of anti-
DNA or rheumatoid factor in their circulation. The clearance
defect is however, correlated with disease activity; patients with
the most extensive disease have the most extensive defect in
clearance. Finally, treatment and recovery from the disease leads
to a marked improvement in the clearance defect, although the
abnormality does not disappear in most patients studied over a
period of months. The message I would like you to take from these
studies is that the Fc and C3b receptor function can now be studied
in vivo in patients with a wide variety of diseases, including
parasitic diseases. It is possible to directly assess the Fc-
receptor capacity of the spleen acting as a reticuloendothelial
organ. The time is ripe for the study of these receptors in a
variety of clinical situations and it is possible that we will
obtain much greater understanding of the role of immunological
phenomena in these clinical states.

The second general area I would like to address concerns the
study of the clearance of particles which are not immunologically
neutral, as are red cells. The clearance of isologous erythrocytes
which we have been discussing thusfar can be considered as an

FIGURE 7

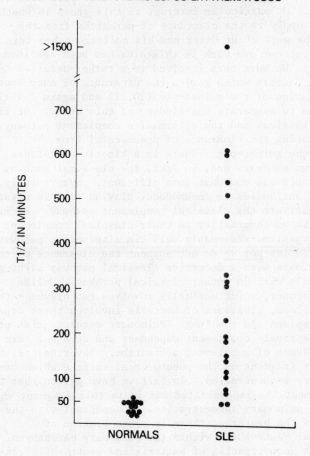

**RETICULOENDOTHELIAL
SYSTEM FC SPECIFIC
CLEARANCE IN
SYSTEMIC LUPUS ERYTHEMATOSUS**

Clearance of IgG anti-Rh coated erythrocytes in man. Shown is clearance in patients with systemic lupus erythematosus as compared to clearance in normals. The major defect in patients with systemic lupus erythematosus is clear.

example of the clearance of a deformable particle which is immunologically inert. These particles must be coated with immunologically-active proteins or protein fragments to be recognized as abnormal as they circulate. It is of great interest to determine whether the general principles which govern the clearance of red cells apply to clearance of other particles from the circulation. Of particular interest to this group is whether such principles apply to the clearance of parasites from the circulation. The work of Dr Oster and his colleagues has been presented. As yet our own work in this area has been confined to bacteria.(20,21) We have been involved in a rather detailed analysis of the factors which govern the clearance of pneumococci from the circulation of the guinea-pig (20,21) and again I will simply have time to enumerate the finding of this study: we find that both the classical and the alternative complement pathways are capable of mediating the clearance of pneumococci from the circulation of the guinea-pig. There is a kinetic difference between these two pathways and, in fact, the classical pathway is somewhat more rapid and somewhat more efficient. Surprisingly, in the guinea-pig, antibodies to pneumococci play no role in clearance other than to activate the classical complement pathway. We find that animals with an abnormality in their classical complement pathway clear organisms reasonably well via alternative pathway function. Antibodies per se do not augment the clearance of these organisms in animals with a defective classical pathway although they do in animals with an intact classical pathway. Unlike erythrocyte clearance, which basically involves two organs - the liver and the spleen, clearance of bacteria involves three organs : the liver, the spleen and the lung. Pulmonary sequestration of pneumococci is entirely complement dependent and does not take place in the absence of complement activation. Nevertheless, there is no complement fragment on the pneumococcal surface which can mediate pulmonary sequestration. In fact we have proven that the complement fragment C5a is generated and it is this fragment which is responsible for pulmonary sequestration. Concomitant with the sequestration of the bacteria, there is sequestration of polymorphonuclear neutrophils within the pulmonary parenchyma. We believe that the sequestration of bacteria and neutrophils can lead to lung damage and in fact have postulated that the "shock lung" syndrome may be a consequence of this interaction. Once again the relevance of these studies to studies of host defence against parasitic disease should be clear. The approaches we have developed can be applied to a variety of microorganisms. I don't pretend to be able to foresee the results of these experiments. Each of the

areas that we investigate will probably lead to their own
complexities and new findings. Nevertheless I believe that an
approach to the problem is available and will lead to greatly
increased understanding of the pathogenesis of parasitic disease.

REFERENCES

1. BIOZZI, G. and STIFFEL, C. The pathophysiology of the
    reticuloendothelial cells of the liver and spleen. Progress in
    Liver Diseases, 2: 166 (1965).

2. BIOZZI, G. ET AL. Exploration of the phagocytic function of
    the reticuloendothelial system with heat denatured human serum
    albumin labeled with I^{131} and application to the measurement
    of liver blood flow in man and in some pathological conditions.
    Journal of Laboratory Clinical Medicine, 51: 230 (1958).

3. MOLLISON, P.L. Blood Transfusion in Clinical Medicine, 5th
    edition (1972).

4. DACIE, J.V. The Haemolytic Anemias: Congenital and Acquired.
    Part II. The Auto-Immune Haemolytic Anemias, 2nd edition (1962).

5. RIFKIND, R.A. Destruction of injured red cells in vivo. American
    Journal of Medicine, 41: 711 (1966).

6. LOBUGLIO, A.F. ET AL. Red cells coated with immunoglobulin G:
    binding and sphering by mononuclear cells in man. Science, 158:
    1582 (1967).

7. HUBER, H. ET AL. Human monocytes: Distinct receptor sites for
    the third component of complement and for immunoglobulin G.
    Science, 162: 1281 (1968).

8. FRANK, M.M. ET AL. Pathophysiology of immune hemolytic anemia.
    Annals of Internal Medicine, 87: 210 (1977).

9. SCHREIBER, A.D. and FRANK, M.M. Role of antibody and complement
    in the immune clearance and destruction of erythrocytes. I.
    In vivo effects of IgG and IgM complement-fixing sites. Journal
    of Clinical Investigation, 51: 575 (1972).

10. SCHREIBER, A.D. and FRANK, M.M. Role of antibody and complement
    in the immune clearance and destruction of erythrocytes II.
    Molecular nature of IgG and IgM complement-fixing sites and
    effects of their interaction with serum. Journal of Clinical
    Investigation, 51: 583 (1972).

11. ATKINSON, J.P. and FRANK, M.M. Studies on the in vivo effects of antibody: Interaction of IgM antibody and complement in the immune clearance and destruction of erythrocytes in man. Journal of Clinical Investigation, 54: 339 (1974).

12. REYNOLDS, H.Y. ET AL. Receptors for immunoglobulin and complement on human alveolar macrophages. Journal of Immunology, 114: 1813 (1975).

13. MANTOVANI, B. ET AL. Phagocytosis of immune complexes by macrophages: Different roles of the macrophage receptor sites for complement C3 and for immunoglobulin IgG. Journal of Experimental Medicine, 135: 780 (1972).

14. ATKINSON, J.P. and FRANK, M.M. Complement-independent clearance of IgG-sensitized erythrocytes: Inhibition by cortisone. Blood 44: 629 (1974).

15. BORSOS, T. ET AL. The C1a fixation and transfer test: Examples of its applicability to the detection and enumeration of antigens and antibodies at cell surfaces. Journal of Immunology 101: 392 (1968).

16. ROSSE, W.F. Fixation of the first component of complement $C^1$1a by human antibodies. Journal of Clinical Investigation, 47: 2430 (1968).

17. EHLENBERGER, A.G. and NUSSENZWEIG, V. The role of membrane receptors for C3b and C3d in phagocytosis. Journal of Experimental Medicine, 145: 357 (1977).

18. JAFFE, C.J. ET AL. Clearance of immune specific particles from the circulation of man: Studies in primary biliary cirrhosis, chronic active hepatitis and alocholic cirrhosis (abstract). Clinical Research, 24: 330A (1976).

19. FRANK, M.M. ET AL. An immuno specific clearance defect in patients with systemic lupus erythematosus (SLE) related to levels of circulating immune complexes (abstract). Clinical Research, 25: 357A (1977).

20. HOSEA, S. and FRANK, M.M. The role of complement in host defense against bacteria (abstract). Journal of Immunology, 120: 1778 (1978).

21. HOSEA, S.W. ET AL. The role of complement in the respiratory distress syndrome (abstract). Clinical Research, 26: 397A (1978)

ACKNOWLEDGEMENTS

Studies of the pathophysiology of red cell clearance in animals were performed in collaboration with Drs Alan Schreiber and John Atkinson. Drs John Atkinson and Charles Jaffe participated in studies of the pathophysiology of erythrocyte clearance in man. The studies of hepatic disease in man were performed in collaboration with Drs Charles Jaffe, E.A. Jones and John Vierling. Studies in lupus erythematosus were performed with Drs Max Hamburger, Paul Plotz and Robert Kimberly. The bacterial clearance studies have been performed in collaboration with Dr Stephen Hosea.

ACKNOWLEDGEMENTS

Studies of the radiophysiology ... ...
were performed ... ... Drs Alan Edmundson and
John Atkinson. Drs John McKim ... Nicholas ... participated in a
study of the dehydrogenation of structures ... ...
The synthesis of new ... ... to new were performed in collaboration
with Drs Charles Tait, P.D. Jones and John Gunther. Studies of
some experiments were performed with Drs New Edmundson ...
Paul Flory and Robert Blanchly. The biochemical experiments ...
have been performed in collaboration with Dr Stephen Hanes.

The Role of the Spleen in the Removal of Senescent Red Cells and the Hypersplenic Syndrome

P.A. Miescher, Division d'Hématologie, Département de
Médecine, Hôpital Cantonal, Geneva, Switzerland

In 1977, Dr Cooper made the following statement: "While
considerable data exist describing the mode of destruction of
abnormal red cells, relatively little is known about normal red cell
ageing and the mode of destruction of senescent red cells".(1)
Ecker, in 1847 (2) and Koelliker in 1849,(3) described cells in the
spleen containing red cells within their cytoplasm. However, these
authors did not at that time recognize the phenomenon of
phagocytosis which was first assessed by Preyer in 1864,(4) being
the first investigator to describe the actual process of
erythrophagocytosis. At that time it was believed that red cells
are first fragmented or lysed before being engulfed. Rous and
Robertson (5) assumed that the senescent red cells become
fragmented before being eliminated from the circulating blood.
Bergenhem and Fahraeus (6) considered lysis of red cells by
subtances similar to lysolecithin to be the main mechanism leading
to the elimination of red cells within the spleen.

With the introduction of radioactive isotopes into
experimental medicine, more insight was gained into the way old red
cells are eliminated from the blood circulation. With this method,
it could first be shown that red cell-survival of normal
erythrocytes is unaffected by the removal of the spleen.(7,8)
However, it was not clear where the old red cells were removed in
splenectomized subjects.

During the years 1955-57, we conducted a series of experiments to better assess the fine mechanism and the site of removal of the senescent red cells. At first, we studied the clearance in vitro of aged red cells in guinea-pigs.(9,10) The rate of elimination of Cr51-labelled red cells proved to be correlated to the amount injected. In contrast, the solution of labelled haemoglobin exhibited a different elimination pattern following an exponential function, with a constant slope if the results were indicated as percentage values of the initial haemoglobin concentration. Since the same labelling was used in the two experiments, the amount of intravascular haemolysis must be insignificant in the elimination of old red cells. The distribution of Cr51 after the injection of aged red cells was similar to the distribution of colloidal particles known to be eliminated by the phagocytic cells of the RES:(11) 95% of the radioactivity could be recovered in the liver.(9) In a second series of experiments we investigated the distribution of Cr51 in animals 40-60 days after autologous red cells had been labelled.(12) Before the organ distribution was studied, the animals were heparinized and carefully perfused with 300 ml of warm Tyrode solution. The input of solution into the jugular vein was balanced with the output of blood from the carotid artery. The distribution of Cr51 in these animals showed the highest amount of Cr51 in the bone marrow (30-75%), followed by the liver (23-49%) and the spleen (6-25%).

From these experiments it was concluded that red cells are eliminated throughout the RES, the distribution depending upon the haemodynamic features. Ehrenstein and Lockner (13) arrived at similar conclusions using different methods. It is interesting to note that rabbits with a shortened red cell survival span showed an increased amount of Cr51 within the spleen.(12) It is possible that these rabbits were not normal but "hypersplenic" (see last paragraph).

One of the main problems is to find out which changes in the senescent red cells are determinant for being trapped by macrophages A number of alterations have been observed such as changes in lipid composition,(14) changes in the amount of sialic acid at the surface of red cells,(15) as well as changes in the enzyme activity.(16,17) For the time being it is not possible to say which change gives the signal for phagocytosis to the macrophages.

With the introduction of isotope methodology it also became possible to better define the so-called hypersplenic condition. Animals rendered hypersplenic as well as hypersplenic patients exhibit different patterns of red-cell sequestration: instead of a

selective sequestration of the senescent red cells as in normal
organisms, the hypersplenic state is characterized by the removal
at random of red cells by the splenic macrophages, i.e. independent
of the age of the red cells. This condition could thus be described
as a state of disturbed haemodynamics with red cells staying too
long in the red pulp and being altered to such an extent that they
are phagocytosed by macrophages. This alteration appears to be
reversible since red cells from hypersplenic patients injected into
normal recipients exhibit a normal survival time.(18)

REFERENCES

1. COOPER, R.A. Destruction of erythrocytes. In: Hematology,
   (Williams, W.J., Beutler, E., Erslev, A.J. and Rundles, R.W.
   eds.) second edition, McGraw-Hill Book Co. p. 216 (1977).

2. ECKER. Ueber die veraenderungen, welche die blutkoerperchen in
   der milz erleiden. Zeitschrift für Rationelle Medizin, 6: 261
   (1847).

3. KOELLIKER, A. Ueber den bau und die verrichtungen der milz.
   Mitteilungen der Zuricher Naturforschungs Gesellschaft,
   vol. 1: 120 (1849).

4. PREYER, W. Ueber amoeboide blutkoerperchen. Virchows Archiv,
   30: 417 (1864).

5. ROUS, P. and ROBERTSON, O.H. The normal fate of erythrocytes.
   The findings in healthy animals. Journal of Experimental
   Medicine, 25: 651 (1917).

6. BERGENHEM, B. and FAHRAEUS, R. Über spontane hämolysinhildung
   im blut, unter besonder berücksichtigung der physiologie der
   milz. Acta Pathologica et Microbiologica Scandinavica, Supp.
   26: 211 (1936).

7. MIESCHER, P.A. The role of the reticuloendothelial system in
   hematoclasia. In: Physiopathology of the RES. Blackwell
   Scientific Publications, Oxford, p. 147 (1957).

8. SINGER, K. and WEISZ, L. The life cycle of the erythrocyte
   after splenectomy and the problem of splenic hemolysis and
   target cell formation. American Journal of Medical Sciences,
   210: 301 (1945).

9. MIESCHER, P.A. Experimentelle studien zum mechanismus der
   erythroklasie im normalen organismus. Klinische Wochenschrift,
   34: 129 (1956).

10. MIESHCER, P.A. Experimentelle untersuchungen ueber die rolle des RES fuer die erythroklasie. In: Radioaktive Isotope in Klinik und Forshung, (Fellinger, K. and Vetter, H. eds.) vol. 2: 43 (1957).

11. HALPERN, B.N. ET AL. Facteurs régissant la fonction phagocytair du système reticulo-endothélial. Revue d'Hématologie, 9: 621 (1954).

12. MIESCHER, P.A. Le mécanisme de l'érythroclasie à l'état normal. Revue d'Hématologie, 11: 248 (1956).

13. EHRENSTEIN, V.G. and LOCKNER, D. Physiological erythrocyte decomposition. Acta Haematologica, 22: 129 (1959).

14. WINTERBOURN, C.C. and BATT, R.D. Lipid composition of human red cells of different ages. Biochimica et Biophysica Acta, 202: 1 (1970).

15. DUROCHER, J.R. ET AL. Role of sialic acid in erythrocyte survival. Blood, 45: 11 (1975).

16. ALLISON, A.C. and BURN, G.P. Enzyme activity as a function of age in the human erythrocyte. British Journal of Haematology, 1: 291 (1955).

17. BERGER, H. ET AL. The reduction of methaemoglobin to haemoglobin in the ageing red cell. Gerontologica Acta, 4: 220 (1960).

18. MIESCHER, P.A. and ROULET, F.C. Das hypersplenic-syndrom. Progress in Surgery, 1: 184 (1961).

DISCUSSION

McGREGOR: Some children experiencing or recovering from severe
P. falciparum malaria possess Coombs (direct) positive erythrocytes.
Such cells usually have C3d attached to them and sometimes also IgG.
Would Dr Frank speculate on the mechanism which would lead to the
attachment of C3d to erythrocytes?

FRANK: I have no direct data on this subject, but it is easy to
raise speculations. Immune complexes may activate complement
components in the region of the cell membrane and such cells can be
coated with C3b in the absence of antibody sensitization. Antibody
molecules may be present on the cells at too low a density to be
detected by the usual techniques and may activate complement;
malaria antigens on the cell surface could conceivably interact
with components of the alternative pathway and activate complement.
There are many possible mechanisms and the techniques are currently
available to evaluate these and work them out.

RUSSELL: While the concept of immune complexes overloading the
receptor mechanisms leading to overflow and disease is appealing –
do you have any evidence relating to an alternative, e.g. patients
with SLE have an innate – possibly genetic – defect of receptor
function?

FRANK: We do not yet know the answer to this very important
question. What is clear is that there are some types of complexes
which are associated with defective receptor functions and other
types of complexes which are not. The most beautiful example of
this is in our studies of patients with Sjögren's syndrome studied
with Max Hamburger and T. Lawley in my laboratory and my rheumatologist
colleagues Dr H. Moutsopoulus and T. Chuced. We have now studied a
large number of patients with this syndrome. 86% of the patients
studied have immune complexes by the Clq binding assay. However,
when clearance studies are performed the patients clearly divide in
two groups. One group has normal clearance, these patients have
glandular disease; disease confined to the face and head. A second
group of patients has Fc receptor clearance defects. These patients
have extra glandular manifestations of their disease – interstitial
nephritis, interstitial pulmonary lung disease, rheumatoid arthritis
etc. This raises the obvious possibility that there are different
types of circulating complexes in these two situations associated
with the different types of disease. However, again we cannot rule
out an intrinsic receptor defecit.

MITCHELL: Can "conditioning" mentioned previously by Dr Oster be explained in terms of the <u>passive</u> nature of erythrocytes in terms of equilibration of membrane lipids with environmental lipids in the depths of the enlarged spleens.

WEISS: I believe so, especially as this conditioning may involve stasis in the red pulp. The red cell is, with the platelets, likely the only "cell" which can't regulate its lipids in its plasma membrane and these lipids rapidly equilibrate with the surrounding plasma or fluid. If additional cholesterol molecules enter the membrane as occurs experimentally in high cholesterol diets or in β hypolypoproteinaemias associated with severe liver disease, the red cell becomes rigid and "spur" shaped, especially in the red pulp of the spleen, and haemolyse more readily. In the red pulp of the spleen, especially with stasis, such changes in lipid compositic of the membrane can occur. Also, as the O_2 tension drops, haemoglobin - normally perocrystalline - becomes more rigid - again phagocytosis and haemolysis. The glucose concentration probably drop with stasis - depriving rbc which are glycolytic - their source of energy. The pH drops and the osmolonty changes. All of these "conditioning" factors make for stasis, congestion, phagocytosis and lysis.

TORRIGIANI: Concerning the blockage of the C3 receptor have you compared what happens in serum positive RA patients who have antiglobulins both of the IgM and IgG classes versus patients with serum negative RA who have practically only IgG antiglobulins?

FRANK: We do not have sufficient data on patients with rheumatoid arthritis to present clear conclusions. We do know, however, that there is no correlation between IgM rheumatoid factor levels and clearance rates.

FAKUNLE: Your comments and findings in groups of patients with various types of liver disease are interesting. Immune complexes have been found in those groups of liver diseases but it is believed that immune complexes have a special role in the pathogenesis of primary biliary cirrhosis. Immune complex-mediated activation of complement with production of breakdown products in primary biliary cirrhosis would explain your findings. Have you got evidence in support of this?

FRANK: Patients with primary biliary cirrhosis have in general normal or elevated levels of complement components. However, they have increased C3 turnover as shown by Potter et al. of your group. Moreover, Tiesburg et al. have reported that they have C3 breakdown products in their circulation. The simplest hypothesis is that these patients have C3 activation ongoing in their circulation with the formation of active C3 fragments. These fragments may compete for and block the C3 receptors on Kupffer cells so that they are unable to remove the C3-coated red cells as they course through the liver.

FRANK: I should like to elaborate on two points: 1) on the statement that circulating macrophages of a special type in patients with haemolytic anaemia use rather than phagocytose cells coated with $\gamma 1$ and $\gamma 3$ IgG. My understanding is that the vast majority of patients with autoimmune haemolytic anaemia have a lytic process which is characterized by extravascular sequestration and destruction of erythrocytes in the cells of the RES. The erythrocytes do not undergo intravascular haemolysis in the usual case although there are certain types of disease and certain types of antibody which are clearly characterized by intravascular haemolysis. I really am not certain that the splenic mechanism responsible for cell destruction is phagocytosis. We can demonstrate phagocytosis of such coated cells in vitro and this serves a simple explanation but other lytic processes may be responsible for splenic "destruction"; 2) on why cells which have interacted with antibody and complement lyse in some cases and not in others. Is this not inconsistent with the one-hit theory? Originally the one-hit theory stated that a site on a red cell which had interacted with all of the components of complement and in whose membrane a hole had been formed would die an osmotic death. I believe that if the complement components are activated at a site on the membrane which is remote from the lipid bilayer no lyses will occur. The cell will be opsonized and may be phagocytosed but it will not lyse. Thus the stereochemistry of complement activation is of great importance and the failure of lysis is not inconsistant with the one hit theory.

MILLER: Perhaps Professor Morris might comment finally on the role of the spleen in the immune response to tumours.

MORRIS: It would be inappropriate to finish this meeting without saying something about the role of the spleen in immunological responses to tumours and tissue allografts, some of which may be relevant to parasitic infections. The role of the spleen in the immune response to tumours has always attracted a certain amount of interest, for it is well known that it is a relatively rare site for metastases compared to the rest of the lymphoid system, although it is commonly involved in primary tumours of the lymphoid system such as Hodgkin's disease.

Splenectomy has been shown in certain experimental models to reduce the rate of growth of transplanted tumours, increase survival times of animals with transplanted tumours, and increase the incidence of spontaneous regression of transplanted tumours. This protective effect of splenectomy can be restored with adoptive transfer of normal spleen cells, suggesting that a subpopulation of spleen cells produces some tumour-enhancing effect. In the clinic we sometimes see a parallel to these experimental findings when we remove the spleen in Hodgkin's disease as part of a staging laparotomy or for hypersplenism. Not infrequently a patient improves dramatically and other tumour masses, such as involved nodes in the neck, shrink within days of splenectomy despite the presence of widespread disease. The only demonstrable immunological changes in such patients to date have been a fall in IgA and IgG levels after splenectomy. On the other hand, there is no evidence to suggest that splenectomy in normal people for trauma leads to an increased incidence of tumours in later life. Similar findings to those in the tumour experiments have been demonstrated in a skin graft model by Streilein's group, where female mice rejected male skin grafts in an accelerated manner after splenectomy.

However, in contrast to this experimental work are the surprising findings that I alluded to earlier, namely that splenectomy before, and indeed for a few days after, transplantation of a kidney in the rat leads to prolonged survival of the animal with the transplanted kidney. This effect is in general seen only in relatively weak models of histoincompatibility, at least in our hands. Furthermore this effect is not abolished by normal spleen cells and can be augmented by the concurrent use of ALS. This latter finding is not surprising in view of the inevitable reduction in the cytotoxic antibody response to donor histocompatibility antigens that splenectomy produces, even in very strong models of histoincompatibility. Thus, in this model, the spleen itself as an intact organ appears to be quite important, in contrast to the tumour models in which normal spleen cells restore the splenectomized animal to normal.

List of Contributors

Dr A.C. ALLISON, Division of Cell Pathology, Clinical Research Centre, Watford Road, Harrow, Middlesex HA1 3UJ, England.

Dr J.R. BATTISTO, Department of Immunology, Research Division, Cleveland Clinic, Cleveland, Ohio 44106, USA.

Professor Z. BRENER, Department of Parasitology, University of Minas Gerais and Centro de Pesquisas René Rachou, FIOCRUZ, Belo Horizonte - MG, Brazil.

Dr Christine CLAYTON, Division of Immunology, National Institute of Medical Research, London NW7 1AA, England.

Professor S. COHEN, Department of Chemical Pathology, Guy's Hospital Medical School, London SE1 9RT, England.

Dr G.G. CRANE, Division of Haematology, Repatriation General Hospital, Concord 2139, NSW, Australia.

Dr Aba S. DAVID-WEST, Department of Haematology, University College Hospital, Ibadan, Nigeria.

Dr Y.M. FAKUNLE, Ahmadu Bello University, Zaria, Nigeria.

Professor W.L. FORD, Department of Pathology, University of Manchester, Manchester M13 9PT, England.

Dr M.M. FRANK, National Institute of Allergy and Infectious Diseases, National Institutes of Health, Bethesda, Maryland 20014, USA.

Dr B.M. GREENWOOD, Ahmadu Bello University, Zaria, Nigeria.

Professor J.H. HUMPHREY, Department of Immunology, Royal Postgraduate Medical School, Hammersmith Hospital, London W12, England.

Dr Jean LANGHORNE, Department of Chemical Pathology, Guy's Hospital Medical School, London SE1 9RT, England.

Dr I. McGREGOR, Medical Research Council Laboratory, Fajara, P.O.B. 273, Banjul, Gambia.

Professor P.A. MIESCHER, Département de Médecine, Division d'Hématologie, Hôpital cantonal, 1211 Geneva 4, Switzerland.

Dr J.F.A.P. MILLER, Walter and Eliza Hall Institute of Medical Research, Post Office, Royal Melbourne Hospital, Melbourne, Victoria 3050, Australia.

Dr G.F. MITCHELL, Laboratory of Immunoparasitology, The Walter and Eliza Hall Institute of Medical Research, Post Office, Royal Melbourne Hospital, Melbourne, Victoria 3050, Australia.

Professor P.J. MORRIS, Nuffield Department of Surgery, Radcliffe Infirmary, Oxford OX2 6HE, England.

Dr Nadia NOGUEIRA, Department of Cellular Physiology and Immunology, The Rockefeller University, New York, NY 10021, USA.

Professor Ruth NUSSENZWEIG, Division of Parasitology, New York University Medical Center, New York, NY 10016, USA.

Dr C.N. OSTER, Laboratory of Parasitic Diseases, National Institutes of Health, Bethesda, Maryland 20014, USA.

Professor N.N. OZERETSKOVSKAYA, E.I. Martsinovsky Institute of Medical Parasitology and Tropical Medicine, Moscow, USSR.

Dr A. POLTERA, WHO Immunology Research and Training Centre, Centre de Transfusion, Hôpital cantonal, 1211 Geneva 4, Switzerland.

Dr T.S. RUSSELL, Malaria, World Health Organization, 1211 Geneva 27, Switzerland.

Dr J. SPRENT, Immunobiology Unit, Department of Pathology, University of Pennsylvania, School of Medicine, Philadelphia, Pennsylvania 19104, USA.

Dr G. TORRIGIANI, Immunology, World Health Organization, 1211 Geneva 27, Switzerland.

Dr W. TRAGER, The Rockefeller University, New York, NY 10021, USA.

Dr L.P. WEISS, Department of Animal Biology, School of Veterinary Medicine, University of Pennsylvania, Philadelphia, Pennsylvania 19104, USA.

Dr M. ZEMBALA, Division of Microbiology and Clinical Immunology, Institute of Paediatrics, Medical Academy in Krakow, Krakow, Poland.